MULTIVALENT FUNCTIONS

BY

W. K. HAYMAN

SC.D., F.R.S.

Professor of Pure Mathematics
in the University of London,
Imperial College

CAMBRIDGE
AT THE UNIVERSITY PRESS
1958

PUBLISHED BY
THE SYNDICS OF THE CAMBRIDGE UNIVERSITY PRESS

Bentley House, 200 Euston Road, London, N.W. 1
American Branch: 32 East 57th Street, New York 22, N.Y.

CAMBRIDGE UNIVERSITY PRESS
1958

Printed in Great Britain at the University Press, Cambridge
(Brooke Crutchley, University Printer)

CONTENTS

Preface   *page* vii

1. Elementary bounds for univalent functions   1

2. The growth of finitely mean valent functions   17

3. Means and coefficients   41

4. Symmetrization   58

5. Circumferentially mean p-valent functions   94

6. The Löwner theory   117

Bibliography   145

Index   149

PREFACE

Suppose that we are given a function $f(z)$ regular in the unit circle, and that the equation $f(z) = w$ has there

(a) never more than one solution;

(b) never more than p solutions; or

(c) at most p solutions in some average sense,

as w moves over the open plane. Then $f(z)$ is respectively univalent, p-valent or mean p-valent in $|z| < 1$.

It is the aim of this book to study what we can say about the growth of such functions $f(z)$ and, in particular, to obtain bounds for the modulus and coefficients of $f(z)$ and related quantities. Thus our aim is entirely quantitative in character.

The univalent functions represent the classical case of this theory, and we shall study them in Chapters 1 and 6. By and large the methods of these chapters do not generalize to p-valent or mean p-valent functions. The latter two are studied in Chapters 2, 3 and 5. The theory of symmetrization is developed in Chapter 4, both for its applications to Chapter 5 and for its intrinsic interest. This chapter could reasonably be read by itself. Chapter 6 could be read immediately after Chapter 1 by the student interested mainly in univalent functions. Otherwise the chapters depend on preceding work.

The majority of the material here collected has not, to my knowledge, appeared in book form before, and some of it is quite new. I am, however, extremely indebted to G. M. Golusin's tract [2] for the contents of each of Chapters 1 and 6. Montel [1] should also be mentioned, though his approach is rather different from mine. In a tract of this size it is not possible to be exhaustive. Thus I have not been able to find space for Schiffer's variational method, nor Jenkins's theory of modules, both of which have recently scored fine successes in the general field of this book, but I have tried to give references to these results as far as possible. The variational method is developed in Schaeffer and Spencer[2] and Jenkins will cover his theory in a forthcoming tract in the Ergebnisse Series.

The book does demand certain previous knowledge of function theory. Most of this would be contained in the undergraduate course as given, for instance, in Cambridge. When something further is required I have tried to give references to Ahlfors[1] or Titchmarsh[1] where the results in question can be found. Apart from such references it has been my aim to give detailed proofs of all the theorems. In several cases there is a rather difficult key theorem, from which a number of applications follow fairly simply. In such a case, the reader may omit the proof of the basic theorem on a first reading, until convinced of its value by the application.

Finally I should like to thank all those persons who have helped me with this book, and in particular, Professor Kennedy, Dr Smithies, Dr Kövari, Dr Clunie and Mr Axtell for much patient criticism in the proof stage and earlier, and Mr Barry, who kindly prepared the index for me. I am also grateful to the editors for allowing me to publish this book in the Cambridge Tracts series, and to the Cambridge University Press for their patience and helpfulness during all the stages of the preparation of this work.

W. K. H.

LONDON
January 1958

CHAPTER 1

ELEMENTARY BOUNDS FOR UNIVALENT FUNCTIONS

1.0. Introduction. A domain is an open connected set. A function $f(z)$ regular in a domain D is said to be *univalent* in D, if $w = f(z)$ assumes different values w for different z in D. In this case the equation $f(z) = w$ has at most one root in D for any complex w. Such functions map D (1, 1) conformally onto a domain in the w plane.

In this chapter we shall obtain some classical results, which give limits for the growth of functions univalent in the unit circle $|z| < 1$. Most of the rest of this tract will aim at generalizing these theorems by proving corresponding results for p-valent functions, i.e. those for which the equation $f(z) = w$ has at most p roots in D, either for every complex w, or in some average sense as w moves over the plane.

If $f(z) = \sum_0^\infty a_n z^n$ is univalent in $|z| < 1$, then so are $f(z) - a_0$ and $(f(z) - a_0)/a_1$, since $a_1 = f'(0) \neq 0$. In fact if a_1 were zero, $f(z)$ would take all values sufficiently near $w = a_0$ at least twice. We thus study the normalized class $\mathfrak{S}$ of functions

$$w = f(z) = z + a_2 z^2 + \dots$$

univalent in $|z| < 1$.

The two equivalent basic results here are due to Bieberbach [1] and state that for $f(z) \in \mathfrak{S}$,† $|a_2| \leqslant 2$, and that $f(z)$ assumes every value w such that $|w| < \frac{1}{4}$. This latter theorem had been previously proved with a smaller absolute constant by Koebe [1]. The results of Bieberbach are best possible. We shall first prove them and then develop some of their main consequences.

---

† This symbolic statement stands for '$f(z)$ belongs to the class $\mathfrak{S}$'.

1.1. We have

THEOREM 1.1. *Suppose that* $f(z) \in \mathfrak{S}$. *Then* $|a_2| \leqslant 2$, *with equality only for the functions*

$$f_\theta(z) = \frac{z}{(1 - z\,e^{i\theta})^2} = z + 2z^2\,e^{i\theta} + 3z^3\,e^{2i\theta} + \dots \qquad (1.1)$$

We need the following preliminary result:

LEMMA 1.1. *Suppose that* $w = f(z) = \sum\limits_{n=-\infty}^{+\infty} a_n z^n$ *is regular in a domain containing* $|z| = r$, *and that the image of* $|z| = r$ *by* $f(z)$ *is a simple closed curve* $J(r)$, *described once. Then the area* $A(r)$ *enclosed by* $J(r)$ *is* $\pi \left| \sum\limits_{n=-\infty}^{+\infty} n\,|a_n|^2\,r^{2n} \right|$.

We write $w = f(r\,e^{i\theta}) = u(\theta) + iv(\theta)$, where

$$u(\theta) = \frac{1}{2} \sum_{-\infty}^{+\infty} [a_n\,e^{in\theta} + \bar{a}_n\,e^{-in\theta}]\,r^n,$$

$$v(\theta) = \frac{1}{2i} \sum_{-\infty}^{+\infty} [a_n\,e^{in\theta} - \bar{a}_n\,e^{-in\theta}]\,r^n.$$

Thus

$$A(r) = \left| \int_0^{2\pi} u\,\frac{dv}{d\theta}\,d\theta \right|$$

$$= \frac{1}{4} \left| \int_0^{2\pi} \left[\sum_{m=-\infty}^{+\infty} r^m (a_m\,e^{im\theta} + \bar{a}_m\,e^{-im\theta}) \right] \right.$$

$$\times \left. \left[\sum_{n=-\infty}^{+\infty} nr^n (a_n\,e^{in\theta} + \bar{a}_n\,e^{-in\theta}) \right] d\theta \right|$$

$$= \left| \frac{\pi}{2} \sum_{n=-\infty}^{+\infty} [a_n(-na_{-n} + nr^{2n}\bar{a}_n) + \bar{a}_n(nr^{2n}a_n - n\bar{a}_{-n})] \right|$$

$$= \pi \left| \sum_{-\infty}^{+\infty} n\,|a_n|^2\,r^{2n} \right|,$$

since $\Sigma na_n a_{-n} = \Sigma n\bar{a}_n \bar{a}_{-n} = 0$, as we see on replacing n by $-n$ in the summation. Thus the lemma is proved.

Suppose now that

$$w = f(z) = z + a_2 z^2 + \ldots \in \mathfrak{S}.$$

Then so does $F(z) = [f(z^2)]^{\frac{1}{2}} = z + \frac{1}{2} a_2 z^3 + \ldots$. In fact $f(z^2)$ does not vanish except at $z = 0$, where it has a double zero, and if $F(z_1) = F(z_2)$, then $f(z_1^2) = f(z_2^2)$, and so $z_1^2 = z_2^2$, i.e. $z_1 = \mp z_2$. But $F(z)$ is an odd function, so that $z_1 = -z_2$ gives $F(z_1) = -F(z_2)$. Hence we must have $z_1 = z_2$. Also since $f(z^2)$ has only a single zero of order two, $F(z)$ is regular. Therefore $F(z)$ is univalent.

Next let

$$g(z) = \frac{1}{F(z)} = \frac{1}{z} - \frac{1}{2} a_2 z + \ldots = \frac{1}{z} + \sum_{n=1}^{\infty} b_n z^n.$$

Then $g(z)$ is univalent in $0 < |z| < 1$, and so the image of $|z| = r$ by $g(z)$ is a simple closed curve for $0 < r < 1$. Hence by Lemma 1·1

$$-\frac{1}{r^2} + \sum_{n=1}^{\infty} n |b_n|^2 r^{2n} = \mp \frac{A(r)}{\pi}$$

does not vanish for $0 < r < 1$. The left-hand side is clearly negative for small positive r and so for $0 < r < 1$. Making $r \to 1$ we deduce

$$\sum_{n=1}^{\infty} n |b_n|^2 \leqslant 1.$$

Thus we have $|b_1| = \frac{1}{2} |a_2| \leqslant 1$, and equality is possible only if $b_n = 0$ $(n > 1)$, and in this case

$$g(z) = \frac{1}{z} - z e^{i\theta}, \quad F(z) = \frac{z}{1 - z^2 e^{i\theta}}, \quad f(z) = \frac{z}{(1 - z e^{i\theta})^2}.$$

This proves Theorem 1.1.

We deduce immediately

THEOREM 1.2. *Suppose that $f(z) \in \mathfrak{S}$ and that $f(z) \neq w$ in $|z| < 1$. Then $|w| \geqslant \frac{1}{4}$. Equality is possible only if $f(z)$ is given by (1.1) and $w = -\frac{1}{4} e^{-i\theta}$.*

Since $f(z) \neq w$

$$\frac{wf(z)}{w - f(z)} = z + \left(a_2 + \frac{1}{w}\right) z^2 + \ldots \in \mathfrak{S}.$$

Thus Theorem 1.1 gives

$$\left| a_2 + \frac{1}{w} \right| \leqslant 2, \quad \left| \frac{1}{w} \right| \leqslant 2 + |a_2| \leqslant 4, \quad |w| \geqslant \frac{1}{4},$$

as required. Equality is possible only if $a_2 = 2e^{i\theta}$, $w^{-1} = -4e^{i\theta}$, and then $f(z)$ must be given by $f_\theta(z)$ in (1·1).

We note finally that the function

$$f_0(z) = \frac{z}{(1-z)^2} = \frac{1}{4}\left(\frac{1+z}{1-z}\right)^2 - \frac{1}{4}$$

maps $|z| < 1$ $(1, 1)$ conformally onto the w plane cut from $-\frac{1}{4}$ to $-\infty$ along the negative real axis. Thus $f_0(z) \in \mathfrak{S}$ and $f_0(z) \neq -\frac{1}{4}$ in $|z| < 1$. Hence the functions $f_\theta(z) = e^{-i\theta} f_0(z\,e^{i\theta})$ of (1.1) also belong to $\mathfrak{S}$ and $f_\theta(z) \neq -\frac{1}{4}e^{-i\theta}$. Thus Theorems 1.1 and 1.2 are best possible.

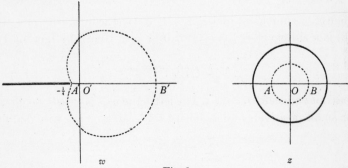

Fig. 1

We shall see that these functions $f_\theta(z)$ are extreme in $\mathfrak{S}$ for a variety of other problems also. It is one of the deepest un-solved problems in the theory whether they are extreme for the nth coefficient, i.e. whether $|a_n| \leqslant n$ holds for all $n > 1$ and $f(z) \in \mathfrak{S}$. Theorem 1.1 proves this for $n = 2$. We shall give the proof when $n = 3$, which is due to Löwner[2], in Chapter 6. Just recently Garabedian and Schiffer [2] have proved the result for $n = 4$, but the general case remains open.

1.2. Elementary growth and distortion theorems. We can develop an interesting further group of inequalities as a direct consequence of Theorem 1.1.

THEOREM 1.3. *Suppose that* $f(z) \in \mathfrak{S}$. *Then we have*† *for* $|z| = r$ $(0 < r < 1)$

$$\frac{r}{(1+r)^2} \leqslant |f(z)| \leqslant \frac{r}{(1-r)^2}, \tag{1·2}$$

† Bieberbach[1], Gronwall[1], Szegö[1].

$$\frac{1-r}{(1+r)^3} \leqslant |f'(z)| \leqslant \frac{1+r}{(1-r)^3}, \tag{1.3}$$

$$\frac{1-r}{r(1+r)} \leqslant \left|\frac{f'(z)}{f(z)}\right| \leqslant \frac{1+r}{r(1-r)}. \tag{1.4}$$

Equality holds in all cases only for the functions $f_\theta(z)$ of (1.1).

We assume $|z_0| < 1$, and set

$$\phi(z) = f\left(\frac{z_0 + z}{1 + \bar{z}_0 z}\right) = b_0 + b_1 z + b_2 z^2 + \dots \tag{1.5}$$

Then clearly $\phi(z)$ is univalent in $|z| < 1$. Further

$$b_0 = f(z_0), \quad b_1 = \phi'(0) = (1 - |z_0|^2) f'(z_0),$$

$$b_2 = \tfrac{1}{2}\phi''(0) = \tfrac{1}{2}(1 - |z_0|^2)^2 f''(z_0) - \bar{z}_0(1 - |z_0|^2) f'(z_0).$$

We apply Theorem 1.1 to $(\phi(z) - b_0)/b_1 \in \mathfrak{S}$ and obtain $|b_2| \leqslant 2|b_1|$, i.e.

$$|f''(z_0)(1 - |z_0|^2)^2 - 2\bar{z}_0 f'(z_0)(1 - |z_0|^2)| \leqslant 4(1 - |z_0|^2)|f'(z_0)|.$$

Writing $z_0 = \rho e^{i\theta}$ we deduce

$$\left| z_0 \frac{f''(z_0)}{f'(z_0)} - \frac{2\rho^2}{1 - \rho^2} \right| \leqslant \frac{4\rho}{1 - \rho^2}. \tag{1.6}$$

Since

$$\frac{\partial}{\partial \rho} \log |f'(\rho e^{i\theta})| = \Re e^{i\theta} \frac{f''(\rho e^{i\theta})}{f'(\rho e^{i\theta})},$$

we obtain at once

$$\frac{2\rho - 4}{1 - \rho^2} \leqslant \frac{\partial}{\partial \rho} \log |f'(\rho e^{i\theta})| \leqslant \frac{2\rho + 4}{1 - \rho^2}.$$

On integrating this from 0 to r with respect to ρ, we deduce (1.3).

We deduce immediately that

$$|f(r e^{i\theta})| \leqslant \int_0^r |f'(\rho e^{i\theta})| \, d\rho \leqslant \int_0^r \frac{1 + \rho}{(1 - \rho)^3} \, d\rho = \frac{r}{(1 - r)^2},$$

and this is the right-hand inequality in (1.2). To obtain the lower bound for $|f(r e^{i\theta})|$, we assume without loss in generality, that $f(r e^{i\theta}) = R e^{i\phi}$, where $R < \tfrac{1}{4}$, since otherwise there is nothing to prove. It then follows from Theorem 1.2 that the straight line segment λ from 0 to $R\,e^{i\phi}$ lies entirely in the image of $|z|<1$ by $f(z)$. Hence λ corresponds to a path l in $|z|<1$, which joins $z=0$ to $r\,e^{i\theta}$. Thus if $t=|z|$ we deduce from (1.3)

$$R=\int_{\lambda}|\,dw\,|=\int_{l}\left|\frac{dw}{dz}\right||\,dz\,|\geqslant\int_{l}\frac{(1-t)}{(1+t)^{3}}\,dt=\frac{r}{(1+r)^{2}},$$

and this completes the proof of (1.2).

Finally, we apply (1.2) to $[\phi(z)-b_{0}]/b_{1}$, where $\phi(z)$ is defined by (1·5). This gives

$$|\,b_{1}\,|\frac{|\,z\,|}{(1+|\,z\,|)^{2}}\leqslant\left|f\left(\frac{z_{0}+z}{1+\bar{z}_{0}z}\right)-f(z_{0})\right|\leqslant|\,b_{1}\,|\frac{|\,z\,|}{(1-|\,z\,|)^{2}}.$$

Putting $z=-z_{0}$, $b_{1}=(1-|\,z_{0}\,|^{2})\,f'(z_{0})$, we deduce (1.4).

It is easily seen that the functions $f_{\theta}(z)$ of (1.1) yield equality in the right-hand inequalities of (1.2)–(1.4) when $z=r\,e^{-i\theta}$, and in the left-hand inequalities if $z=-r\,e^{-i\theta}$. On noting that by Theorem 1.1 equality is possible in (1.6) and hence in the subsequent inequalities only if $\phi(z)$ reduces to one of the functions $f_{\theta}(z)$, we easily see that no other functions can give equality in Theorem 1.3.

1.2.1. We develop now another proof of some of the results in Theorem 1.3, which is based on Theorem 1.2, rather than 1.1. Once Theorem 1.2 has been extended to more general classes of functions, the present proof will also generalize. To make this evident we introduce the following:

DEFINITION. *Let* $f(z)=z+a_{2}z^{2}+\dots$ *be regular in* $|z|<1$. *We shall say that* $f(z)\in\mathfrak{S}_{0}$ *if, given any complex* z_{0}, *with* $|\,z_{0}\,|<1$, *and any function* $\omega(\zeta)$ *univalent and satisfying* $|\,\omega(\zeta)\,|<1$, $\omega(\zeta)\neq z_{0}$ *in* $|\,\zeta\,|<1$, *we have for* $\phi(\zeta)=f[\omega(\zeta)]$

$$|\,\phi'(0)\,|\leqslant 4(|\,\phi(0)\,|+|\,f(z_{0})\,|).$$

We note that $\mathfrak{S}$ is a subclass of $\mathfrak{S}_{0}$. For if $f(z)\in\mathfrak{S}$, $\phi(\zeta)=f[\omega(\zeta)]$ is univalent in $|\,\zeta\,|<1$, and $\phi(\zeta)\neq f(z_{0})=w_{0}$ say. We write

$\phi(\zeta) = b_0 + b_1\zeta + \dots$, and apply Theorem 1.2 to $(\phi(\zeta) - b_0)/b_1 \in \mathfrak{S}$, which never takes the value $(w_0 - b_0)/b_1$. Thus

$$\left| \frac{w_0 - b_0}{b_1} \right| \geq \frac{1}{4}, \quad |b_1| = |\phi'(0)| \leq 4 |w_0 - b_0| \leq 4[|f(z_0)| + |\phi(0)|],$$

and so $f(z) \in \mathfrak{S}_0$.

We shall see in Chapter 5 that the class $\mathfrak{S}_0$ is effectively a good deal larger than $\mathfrak{S}$, and so the results of Theorems 1.4 and 1.5 will apply to a significantly more general class of functions than the univalent ones.

THEOREM 1.4. *Suppose that* $f(z) = z + a_2 z^2 + \dots \in \mathfrak{S}_0$. *Then*

$$|a_2| \leq 2. \tag{1.7}$$

Further, we have for $|z| = r \ (0 < r < 1)$

$$\frac{r}{(1+r)^2} \leq |f(z)| \leq \frac{r}{(1-r)^2}, \tag{1.8}$$

$$|f'(z)| \leq \frac{1+r}{r(1-r)} |f(z)| \leq \frac{1+r}{(1-r)^3}. \tag{1.9}$$

Finally, the equation $f(z) = w$ *has exactly one root in* $|z| < 1$ *if* $|w| < \frac{1}{4}$.

In return for our greater generality we have lost only the left inequalities in (1.3) and (1.4). This is inevitable, since the derivatives of functions of $\mathfrak{S}_0$ may well vanish in $|z| < 1$.

To prove Theorem 1.4, put

$$\frac{z}{(1-z)^2} = Z = \frac{4d\zeta}{(1-\zeta)^2},$$

where $d = \dfrac{r}{(1+r)^2}$ for some fixed r satisfying $0 < r < 1$. Then $|z| < 1$ cut from $-r$ to -1 along the negative real axis is mapped $(1,1)$ conformally onto the Z plane cut from $-d$ to $-\infty$ along the negative real axis and so onto $|\zeta| < 1$. Thus if we write $z = \omega(\zeta)$, $\phi(\zeta) = f[\omega(\zeta)]$, then $\omega(\zeta)$ is univalent, $\omega(\zeta) \neq -r$ in $|\zeta| < 1$, and so, since $f(z) \in \mathfrak{S}_0$, we have

$$|\phi'(0)| \leq 4\{|\phi(0)| + |f(-r)|\},$$

i.e. $\qquad\qquad 4d |f'(0)| \leq 4 |f(-r)|.$

Since $f'(0) = 1$, this gives $|f(-r)| \geqslant d$, and on applying the argument to $e^{-i\theta} f(z\, e^{i\theta})$, which belongs to $\mathfrak{S}_0$ if $f(z) \in \mathfrak{S}_0$, we have the left inequality of (1.8).

It follows immediately that $f(z) \neq 0$ in $|z| < 1$ except at $z = 0$. Next it follows from Rouché's theorem† that if $|w| < r(1+r)^{-2}$, $f(z)$ and $f(z) - w$ have an equal number of zeros in $|z| < r$, i.e. exactly one. Making $r \to 1$, we deduce that for $|w| < \frac{1}{4}$, $f(z) = w$ has exactly one root in $|z| < 1$.

Next choose θ so that $a_2 e^{i\theta} = -|a_2|$. Then as $r \to 0$,

$$|f(r\, e^{i\theta})| = |r + a_2 e^{i\theta} r^2 + O(r^3)| = r - |a_2|\, r^2 + O(r^3),$$

and

$$|f(r\, e^{i\theta})| \geqslant \frac{r}{(1+r)^2} = r - 2r^2 + O(r^3),$$

by the left inequality of (1.8). This gives $|a_2| \leqslant 2$.

It remains to prove the inequalities (1.9). We put

$$Z = \frac{z}{(1-z)^2} = k\left(\frac{1+\zeta}{1-\zeta}\right)^2, \quad \text{where} \quad k = \frac{r}{(1-r)^2}.$$

Here r is a fixed positive number such that $0 < r < 1$. Then $|\zeta| < 1$ corresponds $(1, 1)$ conformally to the Z plane cut along the negative real axis and to $|z| < 1$ cut along the real axis from -1 to 0. We again write $z = \omega(\zeta)$, $\phi(\zeta) = f[\omega(\zeta)]$. Then $\omega(\zeta) \neq 0$ in $|\zeta| < 1$, and since $f(0) = 0$, $f(z) \in \mathfrak{S}_0$, we have

$$|\phi'(0)| = \frac{(1-r)^3}{1+r}\, 4k\, |f'(r)| \leqslant 4\, |\phi(0)| = 4\, |f(r)|.$$

Since $e^{-i\theta} f(z\, e^{i\theta}) \in \mathfrak{S}_0$ also, we deduce

$$|f'(r\, e^{i\theta})| \leqslant \frac{1+r}{k(1-r)^3}\, |f(r\, e^{i\theta})| = \frac{1+r}{r(1-r)}\, |f(r\, e^{i\theta})|,$$

and this is the left inequality of (1.9). Hence

$$\frac{\partial}{\partial r} \log |f(r\, e^{i\theta})| \leqslant \left|\frac{f'(r\, e^{i\theta})}{f(r\, e^{i\theta})}\right| \leqslant \frac{1+r}{r(1-r)}, \tag{1.10}$$

and integrating this from r_1 to r_2, where $0 < r_1 < r_2 < 1$, we deduce

$$\log\left|\frac{f(r_2\, e^{i\theta})}{f(r_1\, e^{i\theta})}\right| \leqslant \int_{r_1}^{r_2} \frac{(1+r)\, dr}{r(1-r)} = \log\left[\frac{(1-r_1)^2\, r_2}{r_1(1-r_2)^2}\right],$$

† See e.g. Ahlfors [2], p. 124.

or

$$\frac{(1-r_2)^2}{r_2}\,|f(r_2\,e^{i\theta})| \leqslant \frac{(1-r_1)^2}{r_1}\,|f(r_1\,e^{i\theta})|. \qquad (1.11)$$

Making $r_1 \to 0$ in this we deduce the right-hand inequality of (1.8) and (1.9) with $r = r_2$. This completes the proof of Theorem 1.4.

From the point of view of later applications it is worth while to note the following consequences of (1.9):

THEOREM 1.5. *Suppose that* $f(z) = z + a_2 z^2 + \ldots \in \mathfrak{S}_0$ *and set*

$$M(r,f) = \max_{|z|=r}|f(z)| \qquad (0 < r < 1).$$

Then unless $f(z) = f_\theta(z) = z(1-z\,e^{i\theta})^{-2}$, $(1-r)^2\,r^{-1}M(r,f)$ *decreases strictly with increasing* r $(0 < r < 1)$, *and so tends to* $\alpha < 1$ *as* $r \to 1$. *Hence the upper bounds for* $|f(z)|$, $|f'(z)|$ *given by (1.8) and (1.9) respectively are attained only by the functions* $f_\theta(z)$.

To prove Theorem 1.5 note that equality can hold in (1.11) only if equality holds in both the inequalities of (1.10) for $r_1 < r < r_2$. This gives

$$\Re\,e^{i\theta}\frac{f'(r\,e^{i\theta})}{f(r\,e^{i\theta})} = \frac{1+r}{r(1-r)},$$

and so

$$\Im\,e^{i\theta}\frac{f'(r\,e^{i\theta})}{f(r\,e^{i\theta})} = 0 \qquad (r_1 < r < r_2),$$

i.e.

$$z\frac{f'(z)}{f(z)} = \frac{1+z\,e^{-i\theta}}{1-z\,e^{-i\theta}}$$

for $z = r\,e^{i\theta}$ $(r_1 < r < r_2)$, and so by analytic continuation throughout $|z| < 1$. In this case $f(z) = f_{-\theta}(z)$.

Otherwise strict inequality holds in (1.11) for $0 < r_1 < r_2 < 1$, $0 \leqslant \theta \leqslant 2\pi$. Choose θ, so that $|f(r_2\,e^{i\theta})| = M(r_2,f)$. Then (1.11) gives

$$\frac{(1-r_2)^2}{r_2}\,M(r_2,f) < \frac{(1-r_1)^2}{r_1}\,|f(r_1\,e^{i\theta})| \leqslant \frac{(1-r_1)^2}{r_1}\,M(r_1,f).$$

Hence unless $f(z) = f_\theta(z)$, $\psi(r) = (1-r)^2\,r^{-1}M(r,f)$ decreases strictly with increasing r $(0 < r < 1)$, and $\psi(r) \leqslant 1$ by (1.9). Thus

$\psi(r) < 1$ $(0 < r < 1)$, so that the upper bounds for $|f(z)|$ in (1.8) and for $|f'(z)|$ in (1.9) are not attained and $\lim_{r \to 1} \psi(r) = \alpha < 1$. This proves Theorem 1.5.

1.3. Means and coefficients. We referred above to Bieberbach's conjecture that $|a_n| \leqslant n$ holds for all $f(z) \in \mathfrak{S}$ and $n \geqslant 2$. In this direction we proceed to prove the inequality $|a_n| < en$ due to Littlewood [1, 3].

THEOREM 1.6. *Suppose that* $f(z) = z + a_2 z^2 + \ldots \in \mathfrak{S}$. *Then*

$$I_1(r, f) = \frac{1}{2\pi} \int_0^{2\pi} |f(r e^{i\theta})| \, d\theta < \frac{r}{1-r} \quad (0 < r < 1), \quad (1.12)$$

and so
$$|a_n| < eI_1\left[\frac{n-1}{n}, f\right] < en \quad (n \geqslant 2). \quad (1.13)$$

As we saw when proving Theorem 1.1,

$$\phi(z) = [f(z^2)]^{\frac{1}{2}} = z + b_3 z^3 + b_5 z^5 + \ldots \in \mathfrak{S},$$

and by Theorem 1.3, applied to $f(z)$, we have $|\phi(z)| \leqslant r/(1-r^2)$ for $|z| \leqslant r$. We note that

$$\frac{1}{2\pi} \int_0^{2\pi} |\phi'(r e^{i\theta})|^2 \, d\theta$$

$$= \frac{1}{2\pi} \int_0^{2\pi} \phi'(r e^{i\theta}) \overline{\phi'(r e^{i\theta})} \, d\theta$$

$$= \frac{1}{2\pi} \int_0^{2\pi} \left(\sum_{n=1}^{\infty} n b_n r^{n-1} e^{i(n-1)\theta} \right) \left(\sum_{m=1}^{\infty} m \bar{b}_m r^{m-1} e^{-i(m-1)\theta} \right) d\theta$$

$$= \sum_{n=1}^{\infty} n^2 |b_n|^2 r^{2n-2}.$$

Thus
$$\pi \sum_{n=1}^{\infty} n |b_n|^2 r^{2n}$$

$$= \int_0^r \rho \, d\rho \int_0^{2\pi} |\phi'(\rho e^{i\theta})|^2 \, d\theta$$

$$= \{\text{area of transform of } |z| < r \text{ by } w = \phi(z)\}$$

$$< \pi \left(\frac{r}{1-r^2} \right)^2.$$

For since $\phi(z)$ is univalent, the area of the transform is at most πR^2, where R is the greatest distance of the transform from $w = 0$.†

Integrating term by term from 0 to r after division by r we obtain

$$\sum_{n=1}^{\infty} |b_n|^2 r^{2n} < \frac{r^2}{1-r^2}.$$

But
$$I_1(r^2, f) = \frac{1}{2\pi} \int_0^{2\pi} |f(r^2 e^{i\theta})| \, d\theta$$

$$= \frac{1}{2\pi} \int_0^{2\pi} |\phi(r e^{i\theta})|^2 \, d\theta$$

$$= \frac{1}{2\pi} \int_0^{2\pi} \phi(r e^{i\theta}) \overline{\phi(r e^{i\theta})} \, d\theta$$

$$= \sum_{n=1}^{\infty} |b_n|^2 r^{2n}.$$

Thus
$$I_1(r^2, f) < \frac{r^2}{1-r^2}.$$

Writing r for r^2 we have (1.12). Again writing $r = 1 - 1/n$ we obtain

$$|a_n| = \frac{1}{2\pi} \left| \int_{|z|=r} \frac{f(z)\,dz}{z^{n+1}} \right| = \frac{1}{2\pi r^n} \left| \int_0^{2\pi} f(r e^{i\theta}) e^{-in\theta} \, d\theta \right|$$

$$\leqslant \frac{1}{r^n} I_1(r, f) \leqslant \frac{1}{r^{n-1}(1-r)} = \left(1 + \frac{1}{n-1}\right)^{n-1} n < en.$$

This gives (1.13) and completes the proof of Theorem 1.6.

The strongest results of the type of Theorem 1.6 known at present are

$$I_1(r, f) < \frac{r}{1-r^2} + 0.55, \quad |a_n| < \tfrac{1}{2}en + 1.51,$$

due to Bazilevič [1]. We shall not be able to prove these, but will show in Chapter 5 that for any fixed $f(z) \in \mathfrak{S}$

$$\frac{|a_n|}{n} \to \alpha \quad \text{as} \quad n \to \infty,$$

where α is the constant of Theorem 1.5. Thus $|a_n| \leqslant n$ for all large n and a fixed $f(z) \in \mathfrak{S}$.

† Equality is clearly excluded here.

1.4. Convex univalent functions.†

It is quite simple to obtain the exact bounds for the coefficients of functions belonging to certain subclasses of $\mathfrak{S}$. In this section we consider functions mapping $|z| < 1$ onto a convex domain D. Such functions we shall call *convex univalent*. A domain D is said to be convex if given w_1 and w_2 in D the straight line segment joining w_1 to w_2 also lies in D. It is then easy to prove by induction that the centre of gravity $\dfrac{1}{n}(w_1 + w_2 + \ldots + w_n)$ of n points w_1 to w_n in D also lies in D.

THEOREM 1.7. *Suppose that* $g(z) = \sum\limits_{n=1}^{\infty} g_n z^n$ *is convex univalent and maps* $|z| < 1$ *onto* D. *Let* $w = h(z) = \sum\limits_{n=1}^{\infty} h_n z^n$ *be regular in* $|z| < 1$ *and assume there only values* w *which lie in* D. *Then* $|h_n| \leqslant |g_1|$ *and in particular* $|g_n| \leqslant |g_1|$ *for* $n \geqslant 1$.

Consider

$$\psi(z) = g^{-1}[h(z)] = \frac{h_1}{g_1} z + \ldots.$$

Then $\psi(z)$ is regular in $|z| < 1$, satisfies $|\psi(z)| < 1$ there and $\psi(0) = 0$. Hence Schwarz's lemma‡ yields $|h_1/g_1| \leqslant 1$, i.e. $|h_1| \leqslant |g_1|$.

Next let $\eta_k \ (1 \leqslant k \leqslant m)$ be the mth roots of unity and consider

$$H(z) = \frac{1}{m} \sum_{k=1}^{m} h(\eta_k z^{1/m}) = h_m z + h_{2m} z^2 + \ldots$$

instead of $h(z)$. Since D is convex and $h(z)$ assumes only values inside D so does $H(z)$, and we deduce $|h_m| \leqslant |g_1| \ (m = 2, 3, \ldots)$. This proves Theorem 1.7.

We deduce that if $g(z) = z + g_2 z^2 + \ldots$ is convex univalent then $|g_n| \leqslant 1$ for $n \geqslant 1$. These inequalities are sharp for every n, as is shown by

$$w = g(z) = \frac{z}{1-z} = z + z^2 + \ldots,$$

which maps $|z| < 1$ onto the half-plane $\Re w > -\frac{1}{2}$.

We can also sharpen Theorems 1.2 and 1.3 for the functions $g(z)$. In each case the function $z/(1-z)$ is extremal.

---

† The results in this section are due to Löwner[1].
‡ See e.g. Ahlfors [2], p. 110, Theorem 13.

THEOREM 1.8. *Suppose that $w = f(z) = z + \dots$ is convex univalent. Then $f(z)$ assumes every value in the disc $|w| < \tfrac{1}{2}$.*

Let D be the image of $|z| < 1$ by $f(z)$ and let $w_0 = r e^{i\theta}$ be a point of smallest modulus lying outside D. By considering $-e^{-i\theta} f(-z e^{i\theta})$ instead of $f(z)$ if necessary, we may suppose that $w_0 = -r$. Then $\Re w > -r$ in D. For suppose, contrary to this, that D contains a point w_1, such that $\Re w_1 < -r$. By the convexity of D that segment, say L, of the line through w_0, w_1, which lies on the side of w_0 opposite to w_1, lies entirely outside D. But L contains points of modulus less than r, giving a contradiction.

Now

$$w = g(z) = \frac{2rz}{1-z} = 2rz + \dots$$

maps $|z| < 1$ onto $D_0 = \{w : \Re w > -r\}$. Since $f(z) = z + \dots$ assumes values in D_0 only, Theorem 1.7 applied with $f(z)$ instead of $h(z)$ gives $2r \geqslant 1$. This proves Theorem 1.8.

We have finally

THEOREM 1.9. *Suppose that $w = f(z) = z + \dots$ is convex univalent. Then we have for $|z| = r$ $(0 < r < 1)$*

$$\frac{r}{1+r} \leqslant |f(z)| \leqslant \frac{r}{1-r},$$

$$\frac{1}{(1+r)^2} \leqslant |f'(z)| \leqslant \frac{1}{(1-r)^2},$$

$$\frac{1}{r(1+r)} \leqslant \left|\frac{f'(z)}{f(z)}\right| \leqslant \frac{1}{r(1-r)}.$$

All these inequalities are sharp with equality holding when $f(z) = z/(1-z)$ and $z = \pm r$. We omit the proofs which are exactly analogous to that of Theorem 1.3, starting with the inequality $|a_2| \leqslant 1$ which follows from Theorem 1.7 instead of $|a_2| \leqslant 2$.

1.5. Typically real functions.† Following Rogosinski we shall call $f(z)$ *typically real* if $f(z)$ is regular in $|z| < 1$ and $f(z)$ is real there if and only if z is real. We have

† Rogosinski[1]. See also Dieudonné[1] and Szász[1] for the present proof.

THEOREM 1.10. *Suppose that $f(z) = z + a_2 z^2 + \ldots$ is typically real. In that case, and in particular if $f(z) \in \mathfrak{S}$ and has real coefficients, we have $|a_n| \leqslant n$ $(n = 2, 3, \ldots)$.*

Let $f(z) = u + iv$ and suppose that $f(z)$ is typically real. Then $f(z)$ is real on the real axis and so has real coefficients. Thus

$$v(r\, e^{i\theta}) = \sum_1^\infty a_n r^n \sin n\theta.$$

Also $v(r\, e^{i\theta})$ has constant sign for $0 < \theta < \pi$ and so

$$|a_n r^n| = \left| \frac{2}{\pi} \int_0^\pi v(r\, e^{i\theta}) \sin n\theta \, d\theta \right|$$

$$\leqslant \frac{2n}{\pi} \int_0^\pi |v(r\, e^{i\theta}) \sin \theta| \, d\theta = nr \quad (0 < r < 1),$$

since $|\sin n\theta| \leqslant n \sin \theta$. Making $r \to 1$, we deduce $|a_n| \leqslant n$.

If $f(z)$ has real coefficients then $f(\bar{z}) = \overline{f(z)}$. Thus if $f(z)$ has real coefficients and is real for some complex z_0, we have $f(z_0) = f(\bar{z}_0)$, which is impossible for a univalent $f(z)$. Thus if $f(z) \in \mathfrak{S}$ and has real coefficients, $f(z)$ is typically real and the above argument applies.

1.6. Starlike univalent functions.†

A domain D in the w plane is said to be starlike (or star-shaped) with respect to a fixed point O in D, if for any point P in D the straight line segment OP also lies in D. If $f(z) \in \mathfrak{S}$ and maps $|z| < 1$ onto a starlike domain with respect to $w = 0$, we shall call $f(z)$ starlike univalent. We have

THEOREM 1.11. *If $f(z) = z + a_2 z^2 + \ldots$ is starlike univalent then*

$$|a_n| \leqslant n \quad (n = 2, 3, \ldots).$$

Let G be the map of $|z| < 1$, G_r the map of $|z| < r$, by $f(z)$. We show first that G_r is starlike with respect to $w = 0$ for $0 < r < 1$.

---

† Nevanlinna[1]. For a generalization of Theorem 1.11 to functions mapping $|z| < 1$ onto a more general class of domains introduced by Kaplan[1] see Reade[1].

In fact if $w \in G$, then $tw \in G$ for $0 < t < 1$, and so the function

$$\psi(z) = f^{-1}[tf(z)]$$

is regular in $|z| < 1$ and satisfies $|\psi(z)| < 1$ there and $\psi(0) = 0$. Thus we have from Schwarz's lemma $|\psi(z)| \leqslant |z|$ in $|z| < 1$.

Suppose now that $w_1 \in G_r$. Then $w_1 = f(z_1)$, where $|z_1| < r$. Hence

$$|f^{-1}(tw_1)| = |\psi(z_1)| \leqslant |z_1| < r.$$

Thus $tw_1 = f(z_2)$ with $|z_2| < r$ and so $tw_1 \in G_r$ for $0 < t < 1$. Hence G_r is starlike.

The boundary of G_r consists of the curve $w = f(r\, e^{i\theta})$ $(0 \leqslant \theta \leqslant 2\pi)$. Since G_r is starlike the radius vector from $w = 0$ to $f(r\, e^{i\theta})$ lies in G_r and so $\arg f(r\, e^{i\theta})$ increases with θ. Thus

$$\frac{\partial}{\partial\theta} \arg f(r\, e^{i\theta}) = \Re\left[\frac{r\, e^{i\theta} f'(r\, e^{i\theta})}{f(r\, e^{i\theta})}\right] \geqslant 0 \quad (0 < r < 1,\ 0 \leqslant \theta \leqslant 2\pi). \quad (1.14)$$

Write now
$$z\frac{f'(z)}{f(z)} = 1 + \sum_{m=1}^{\infty} \alpha_m z^m.$$

Then it follows from (1.14) that

$$w = h(z) = z\frac{f'(z)}{f(z)} - 1 = \sum_{m=1}^{\infty} \alpha_m z^m$$

takes values lying entirely in the convex domain

$$D_0 = \{w : \Re w > -1\}.$$

Also
$$w = g(z) = \frac{2z}{1-z} = 2z + 2z^2 + \dots$$

maps $|z| < 1$ onto D_0. Hence by Theorem 1.7 $|\alpha_m| \leqslant 2$ $(m = 1, 2, \dots)$. Again we have

$$\left(\sum_{n=1}^{\infty} a_n z^n\right)\left(1 + \sum_{m=1}^{\infty} \alpha_m z^m\right) = \sum_1^{\infty} n a_n z^n,$$

and equating coefficients we deduce

Thus
$$n a_n = a_n + \alpha_1 a_{n-1} + \dots + \alpha_{n-1}.$$

$$(n-1)|a_n| = |\alpha_1 a_{n-1} + \dots + \alpha_{n-1}| \leqslant 2(1 + |a_2| + \dots + |a_{n-1}|).$$

If we assume that $|a_k| \leqslant k$ $(k = 1, 2, ..., n-1)$, we deduce

$$(n-1)|a_n| \leqslant n(n-1),$$

and the proof of Theorem 1.11 by induction is complete.

We remark finally that the function

$$\frac{z}{(1-z)^2} = z + 2z^2 + 3z^3 + \dots$$

is both starlike univalent and typically real, so that the inequalities of Theorems 1.10 and 1.11 are sharp.

CHAPTER 2

THE GROWTH OF FINITELY MEAN VALENT FUNCTIONS

2.0. Introduction. We saw in the previous chapter that the assumption that $f(z)$ is univalent in $|z| < 1$ imposes a number of restrictions on the growth of $f(z)$, such as were proved in Theorems 1.3, 1.5 and 1.6. In this and the next chapter we see what results we can obtain under the more general assumption that the equation $f(z) = w$ has at most p roots in some average sense, as w moves over the plane.

In the present chapter we confine ourselves to obtaining bounds for $|f(z)|$. We set

$$M(r, f) = \max_{|z|=r} |f(z)| \quad (0 < r < 1), \tag{2.1}$$

and show that under suitable averaging assumptions

$$M(r, f) = O(1-r)^{-2p} \quad (r \to 1). \tag{2.2}$$

The first result of this type is due to Cartwright[1], who proved (2.2) when p is a positive integer and the equation $f(z) = w$ never has more than p roots in $|z| < 1$ for any w. Such functions are called p-valent. Her method, based on a distortion theorem of Ahlfors[1], was extended by Spencer[3] to the more general case.

The proof of (2.2) is much simpler if we assume $f(z) \neq 0$ in $|z| < 1$. We prove the result first in this special case, before tackling the general one. If $f(z) = a_0 + a_1 z + \dots$ has q zeros in $|z| < 1$, bounds for $M(r, f)$ can be obtained depending on

$$\mu_q = \max_{0 \leqslant \nu \leqslant q} |a_\nu|.$$

This dependence is essential. In fact any polynomial of degree p is p-valent and has at most p zeros in $|z| < 1$, but bounds for $M(r)$ must clearly depend on all the coefficients.

Finally, we show that our method also leads to a number of theorems, which show that $f(z)$ cannot grow too rapidly near several points of $|z| = 1$ simultaneously, and in particular that, if $M(r)$ attains the growth $(1-r)^{-2p}$, then $|f(re^{i\theta})|$ attains this magnitude for a single fixed value of θ, and is quite small for other constant θ as $r \to 1$. Results of this type appear to need the full strength of the methods of this chapter even for univalent functions, and were first proved by Spencer[3].

2.1. A length-area principle. Let $f(z)$ be regular in an open set Δ and let $n(w)$ be the number of roots in Δ of the equation $f(z) = w$. We write

$$p(R) = p(R, \Delta, f) = \frac{1}{2\pi} \int_0^{2\pi} n(R\,e^{i\phi})\,d\phi. \qquad (2.3)$$

The integral exists as a finite or infinite Lebesgue integral, since $n(w) \geqslant 0$.†

The function $p(R)$ will play a dominant role in the sequel. It is clear that $p(R)$, like $n(w)$, increases with expanding domain Δ. Further if $z = t(\zeta)$ maps Δ_1 $(1,1)$ conformally onto Δ, then if $p(R, \Delta_1)$ refers to $f[t(\zeta)]$, we have $p(R, \Delta) = p(R, \Delta_1)$.

We might make the averaging assumption

$$p(R) \leqslant p \quad (0 < R < \infty).$$

This is certainly satisfied if p is a positive integer and $f(z)$ is p-valent in $|z| < 1$. We shall use this assumption in Chapter 5; at present weaker hypotheses will be sufficient.

We shall base our results on a length-area inequality. A result of this type was first proved by Ahlfors[1], and its use in this context is due to Cartwright[1]. Recently a similar result has proved to be a basic tool in conformal and quasi-conformal mapping (see Lelong-Ferrand[1]).

THEOREM 2.1. *Suppose that $f(z)$ is regular in an open set Δ and that $p(R) = p(R, \Delta)$ is defined by (2.3). Let $l(R)$ be the total length of the curves in Δ on which $|f(z)| = R$, and A the area of Δ, supposed finite. Then we have*

$$\int_0^\infty \frac{l(R)^2\,dR}{R\,p(R)} \leqslant 2\pi A,$$

where the integrand is taken be zero if $l(R) = 0$ or $p(R) = +\infty$. In particular, $l(R) < +\infty$ for almost all R for which $p(R) < +\infty$.

† It will be shown in the course of Lemma 5.2 that the sets $n(w) \geqslant K$ are open in the finite plane. Thus $n(R e^{i\phi})$ is certainly measurable.

2.1.1. We first prove Theorem 2.1 when Δ is an open rectangle and $f(z)$ is univalent and not zero in a domain containing Δ and its sides. We say $f(z)$ is univalent and not zero *on* Δ. Then any fixed branch of

$$s(z) = \log f(z) = \sigma + i\tau$$

is also univalent on Δ and maps Δ onto a domain Ω in the s plane. The boundary of Ω is the image of the circumference of Δ by $s(z)$ and is a sectionally analytic Jordan curve. Let θ_σ be the intersection of Ω with the line $\sigma = $ constant. Then θ_σ consists of a finite number of straight-line segments

$$\tau_1 < \tau < \tau_1', \quad \tau_2 < \tau < \tau_2', \quad \ldots .$$

As $s = \sigma + i\tau$ describes θ_σ, $z = x + iy$ describes the set γ_σ in Δ on which $|f(z)| = e^\sigma$.

Since $f(z) = e^{s(z)}$ is univalent, $\tau_\nu' - \tau_\nu \leqslant 2\pi$. On the arc of γ_σ corresponding to the segment $\tau_\nu < \tau < \tau_\nu'$, the equation

$$f(z) = e^{\sigma + i\tau_0}$$

has no roots if $\tau_\nu' < \tau_0 < \tau_\nu + 2\pi$ and one root if $\tau_\nu < \tau_0 < \tau_\nu'$. Thus the corresponding contribution to $p(e^\sigma, \Delta)$ is exactly $(\tau_\nu' - \tau_\nu)/(2\pi)$. Writing $\theta(\sigma)$ for the total measure of θ_σ we see by addition that

$$p(e^\sigma, \Delta) = \frac{1}{2\pi} \Sigma(\tau_\nu' - \tau_\nu) = \frac{\theta(\sigma)}{2\pi}.$$

We now have, using Schwarz's inequality,†

$$l(e^\sigma, \Delta)^2 = \left[\int_{\theta_\sigma} \left| \frac{dz}{ds} \right| d\tau \right]^2 \leqslant \int_{\theta_\sigma} d\tau \int_{\theta_\sigma} \left| \frac{dz}{ds} \right|^2 d\tau$$

$$= \theta(\sigma) \int_{\theta_\sigma} \left| \frac{dz}{ds} \right|^2 d\tau = 2\pi p(e^\sigma, \Delta) \int_{\theta_\sigma} \left| \frac{dz}{ds} \right|^2 d\tau.$$

If σ_1, σ_2 are the lower and upper bounds of σ on Ω this gives

$$\int_{-\infty}^{+\infty} \frac{l(e^\sigma, \Delta)^2}{p(e^\sigma, \Delta)} d\sigma = \int_{\sigma_1}^{\sigma_2} \frac{l(e^\sigma, \Delta)^2}{p(e^\sigma, \Delta)} d\sigma \leqslant 2\pi \int_{\sigma_1}^{\sigma_2} d\sigma \int_{\theta_\sigma} \left| \frac{dz}{ds} \right|^2 d\tau = 2\pi A,$$

† $|\Sigma ab|^2 \leqslant \Sigma|a|^2 \Sigma|b|^2$ or $\left| \int fg\,dx \right|^2 \leqslant \int |f|^2\,dx \int |g|^2\,dx$. For a proof see, for example, Titchmarsh[1], p. 381.

where A is the area of Δ. Writing $e^\sigma = R$ we have

$$\int_0^\infty \frac{l(R, \Delta)^2}{p(R, \Delta)} \frac{dR}{R} \leqslant 2\pi A, \tag{2.4}$$

as required.

2.1.2. To extend our result to the general case we need the theory of the Lebesgue integral. We may assume without loss in generality that $f(z)$ is regular and $f(z) \neq 0$, $f'(z) \neq 0$ in Δ. For otherwise we may consider $f(z)$ in the set Δ_0 consisting of Δ except for the zeros of $f(z)$ and $f'(z)$, without affecting any of the quantities $p(R)$, $l(R)$ and A in Theorem 2.1. Next we express Δ as a countable union of non-overlapping rectangles Δ_ν ($\nu = 1$ to ∞).† At each point z_0 of such a rectangle $f'(z_0) \neq 0$, and so there is a neighbourhood of z_0 in which $f(z)$ is univalent. Thus we may subdivide each Δ_ν into a finite number of smaller rectangles on each of which $f(z)$ is univalent. We therefore assume without loss in generality that $f(z)$ is univalent, $f(z) \neq 0$ on each Δ_ν.

The set of curves γ_R, on which $|f(z)| = R$, meets a side of one of the rectangles Δ_ν in only a finite number of points, unless $|f(z)| \equiv R$ on the side. Thus if we omit the finite or countable set of values of R for which $|f(z)| \equiv R$ on some side of a rectangle Δ_ν, we have

$$p(R, \Delta) = \sum_{\nu=1}^\infty p(R, \Delta_\nu), \tag{2.5}‡$$

$$l(R, \Delta) = \sum_{\nu=1}^\infty l(R, \Delta_\nu), \tag{2.6}§$$

since the sides of the Δ_ν do not then contribute to $p(R, \Delta)$ or $l(R, \Delta)$. Also if A_ν, A are the areas of Δ_ν, Δ respectively, we have∥

$$A = \sum_{\nu=1}^\infty A_\nu.$$

We now set $a_\nu^2 = \dfrac{l(R, \Delta_\nu)^2}{p(R, \Delta_\nu)}, \quad b_\nu^2 = p(R, \Delta_\nu)$

in Schwarz's inequality $(\Sigma a_\nu b_\nu)^2 \leqslant \Sigma a_\nu^2 \Sigma b_\nu^2$ and sum over those

---

† Burkill[1], p. 8.   ‡ This follows from (2.3) and Burkill[1], (B) p. 41.
§ Burkill[1], p. 35.   ∥ Burkill[1], p. 21.

values of ν for which $p(R, \Delta_\nu)$, $l(R, \Delta_\nu)$ are not zero. Using (2.5) and (2.6) we obtain

$$\frac{l(R, \Delta)^2}{p(R, \Delta)} \leqslant \sum_{\nu=1}^{\infty} \frac{l(R, \Delta_\nu)^2}{p(R, \Delta_\nu)},$$

where indeterminate terms are taken to be zero. We now integrate from 0 to ∞ and obtain, using (2.4) applied to each Δ_ν,

$$\int_0^\infty \frac{l(R, \Delta)^2}{p(R, \Delta)} \frac{dR}{R} \leqslant \int_0^\infty \left[\sum_{\nu=1}^{\infty} \frac{l(R, \Delta_\nu)^2}{p(R, \Delta_\nu)} \right] \frac{dR}{R}$$
$$= \sum_{\nu=1}^{\infty} \int_0^\infty \frac{l(R, \Delta_\nu)^2}{p(R, \Delta_\nu)} \frac{dR}{R} \leqslant 2\pi \sum_{\nu=1}^{\infty} A_\nu = 2\pi A.$$

The inversion of integration and summation is justified since the terms are non-negative.† This completes the proof of Theorem 2.1.

2.2. Functions without zeros. We can now prove a basic result relating the growth of a function $f(z)$ with the quantity $p(R)$.

THEOREM 2.2. *Suppose that $f(z)$ is regular and not zero in Δ: $|z| < 1$ and set $|f(0)| = R_1$, $|f(r e^{i\theta})| = R_2$, where $0 < r < 1$, $0 \leqslant \theta < 2\pi$. Then*

$$\left| \int_{R_1}^{R_2} \frac{dR}{Rp(R, \Delta)} \right| \leqslant 2 \left[\log \left(\frac{1+r}{1-r} \right) + \pi \right],$$

where $p(R, \Delta)$ is defined as in (2.3).

Suppose for definiteness that $R_1 < R_2$. If $R_1 > R_2$, so that the integral is negative, the argument is similar. We write

$$\zeta = \xi + i\eta = \log \frac{1 + z e^{-i\theta}}{1 - z e^{-i\theta}}, \qquad (2.7)$$

and put

$$\xi_1 = 0, \quad \xi_2 = \log \frac{1+r}{1-r}.$$

Consider $f[z(\zeta)]$ as a function of ζ in the rectangle

$$D = \{\zeta: \xi_1 - \tfrac{1}{2}\pi < \xi < \xi_2 + \tfrac{1}{2}\pi, \quad |\eta| < \tfrac{1}{2}\pi\}.$$

Then D corresponds to a subset of Δ by the transformation (2.7) and so $p(R, D) \leqslant p(R, \Delta)$. Also $|f|$ changes continuously from R_1 to R_2 as $\zeta = \xi + 0i$ increases from ξ_1 to ξ_2. Thus for $R_1 < R < R_2$

† Burkill[1], (B) p. 41.

there is at least one point on the segment $\xi_1 \leqslant \xi \leqslant \xi_2$ of the real ζ axis, such that $|f(\zeta)| = R$.

We form the level curve γ_R, on which $|f(z)| = R$, through such a point. Then γ_R can be continued as an analytic arc in both directions as long as $f[z(\zeta)]$ continues to be regular and locally univalent, i.e. as long as f is regular and $f' \neq 0$. Excluding the countable set of R for which γ_R meets a zero of f', γ_R either goes to the boundary of D in both directions or else meets itself.

This second case is, however, excluded by the hypothesis that $f(z) \neq 0$. For in this case we should have a subdomain D_0 of D bounded entirely by γ_R on which $|f| = R$. By applying the maximum modulus principle to f and $1/f$ we should deduce that f is constant, a trivial case, which we exclude.

Thus γ_R goes to the boundary in both directions for all but a countable set of values of R, and the segment $\xi_1 \leqslant \xi \leqslant \xi_2$ is distant at least $\frac{1}{2}\pi$ from each point of the boundary of D. Thus we must have $l(R) \geqslant \pi$ for all these values of R and Theorem 2.1 gives

$$\pi^2 \int_{R_1}^{R_2} \frac{dR}{R p(R, \Delta)} \leqslant \int_{R_1}^{R_2} \frac{l(R, D)^2 \, dR}{R p(R, D)} \leqslant 2\pi \cdot \pi [\xi_2 - \xi_1 + \pi].$$

This completes the proof of Theorem 2.2.

2.3. Some averaging assumptions on $p(R)$. If

$$p(R) \leqslant p \quad (0 < R < \infty), \tag{2.8}$$

Theorem 2.2. shows immediately that

$$\frac{1}{p} \left| \log \frac{R_2}{R_1} \right| < 2 \log \left(e^\pi \frac{1+r}{1-r} \right),$$

$$|f(0)| \left(e^{-\pi} \frac{1-r}{1+r} \right)^{2p} < |f(r\,e^{i\theta})| < |f(0)| \left(e^\pi \frac{1+r}{1-r} \right)^{2p},$$

which is the type of result we are aiming at. Following Spencer [2, 3] we show in this section how we may weaken (2.8). We write

$$p(R) = p + h(R),$$

$$W(R) = \int_0^R p(\rho) \, d(\rho^2)$$

$$= p R^2 + \int_0^R h(\rho) \, d(\rho^2) = p R^2 + H(R),$$

say, and have the following inequality:

LEMMA 2.1. *With the above notation*

$$\int_{R_1}^{R_2} \frac{d\rho}{\rho p(\rho)} \geqslant \frac{1}{p}\left\{\log\frac{R_2}{R_1} - \frac{1}{2} - \frac{H(R_2)}{pR_2^2} - \frac{1}{p}\int_{R_1}^{R_2}\frac{H(\rho)\,d\rho}{\rho^3}\right\}.$$

We have

$$\int_{R_1}^{R_2}\frac{d\rho}{\rho p(\rho)} = \int_{R_1}^{R_2}\frac{d\rho}{\rho[p+h(\rho)]} = \int_{R_1}^{R_2}\frac{d\rho}{\rho}\left[\frac{1}{p} - \frac{h(\rho)}{p^2} + \frac{h^2(\rho)}{p^2[p+h(\rho)]}\right]$$

$$\geqslant \frac{1}{p}\log\frac{R_2}{R_1} - \frac{1}{p^2}\int_{R_1}^{R_2}\frac{h(\rho)\,d\rho}{\rho}.$$

Again

$$\int_{R_1}^{R_2}\frac{h(\rho)\,d\rho}{\rho} = \int_{R_1}^{R_2}\frac{dH(\rho)}{2\rho^2} = \frac{H(R_2)}{2R_2^2} - \frac{H(R_1)}{2R_1^2} + \int_{R_1}^{R_2}\frac{H(\rho)\,d\rho}{\rho^3}.$$

Also $W(R)$ is necessarily positive and so $H(R_1) \geqslant -pR_1^2$. Now Lemma 2.1 follows.

Following Spencer we shall in this and the next chapter call a function $f(z)$ *mean p-valent* in a domain Δ, if $f(z)$ is regular in Δ, p is a positive number, and

$$W(R) \leqslant pR^2 \quad (0 < R < \infty).$$

Using the definition (2.3) we see that

$$W(R) = \frac{1}{\pi}\int_0^{2\pi}\int_0^R n(\rho\,e^{i\phi})\,\rho\,d\rho\,d\phi,$$

and so mean p-valency expresses the condition that the average number of roots in Δ of the equation $f(z) = w$ is not greater than p, as w ranges over any disc $|w| < R$. We can now prove

THEOREM 2.3. *Suppose that* $f(z) = a_0 + a_1 z + \ldots$ *is mean p-valent and not zero in* $|z| < 1$. *Then we have for* $|z| = r$ $(0 < r < 1)$

$$\frac{|a_0|}{C}\left(\frac{1-r}{1+r}\right)^{2p} < |f(z)| < |a_0|\,C\left(\frac{1+r}{1-r}\right)^{2p},$$

where $C = e^{\frac{1}{2}+2\pi p}$.

Set $|a_0| = R_1$, $|f(r\,e^{i\theta})| = R_2$ and suppose, for example, $R_2 > R_1$. Then Theorem 2.2 and Lemma 2.1 give

$$2\left\{\log\frac{1+r}{1-r} + \pi\right\} \geqslant \int_{R_1}^{R_2}\frac{dR}{Rp(R)} \geqslant \frac{1}{p}\left\{\log\frac{R_2}{R_1} - \frac{1}{2}\right\},$$

since $H(R) \leqslant 0$ by hypothesis. This gives

$$R_2 < e^{\frac{1}{2}} R_1 \left\{ e^\pi \frac{1+r}{1-r} \right\}^{2p},$$

as required. If $R_1 > R_2$ we obtain similarly

$$R_1 < e^{\frac{1}{2}} R_2 \left\{ e^\pi \frac{1+r}{1-r} \right\}^{2p},$$

and Theorem 2.3 is proved.

2.3.1. We note that in order to prove

$$M(r,f) = O(1-r)^{-2p} \quad (r \to 1),$$

we require that

$$\int_{R_1}^{R_2} \frac{dR}{Rp(R)} > \frac{1}{p} \log \frac{R_2}{R_1} - O(1) \quad (R_2 \to \infty),$$

and for this it would be sufficient to assume that

$$\frac{H(R_2)}{R_2^2} \quad \text{and} \quad \int_{R_1}^{R_2} \frac{H(R) \, dR}{R^3}$$

are bounded as $R_2 \to \infty$. This follows from Lemma 2.1. These latter conditions, unlike the condition for mean p-valency, can be seen to be independent of the choice of origin in the w plane. Another such condition which is still weaker is

$$\varlimsup_{R \to \infty} \frac{W(R)}{R^2} \leqslant p, \quad \text{i.e.} \quad \varlimsup_{R \to \infty} \frac{H(R)}{R^2} \leqslant 0,$$

but this only implies

$$M(r,f) = O(1-r)^{-2p-\epsilon} \quad (r \to 1)$$

for every positive ϵ. Having pointed out the possibility of such generalizations, we shall confine ourselves to mean p-valent functions for the time being, for the sake of simplicity.

The order of magnitude of the bounds in Theorem 2.3 is best possible, as is shown by the function

$$w = f(z) = a_0 \left(\frac{1+z}{1-z} \right)^{2p},$$

which maps $|z| < 1$ onto the sector (possibly self-overlapping) in the w plane given by

$$\left\{w: \left|\arg\frac{w}{a_0}\right| < p\pi \quad (0 < |w| < \infty)\right\},$$

and for which

$$p(R) = p, \quad W(R) = pR^2 \quad (0 < R < \infty).$$

If p is a positive integer $f(z)$ is p-valent. The example raises the question as to whether the constant C may be replaced by 1 in Theorem 2.3. We shall see in Chapter 5 that this is in fact the case, if we replace mean p-valency by the stronger hypothesis (2.8).

2.4. Functions with zeros. We now aim to eliminate the awkward condition $f(z) \neq 0$ from Theorem 2.3. Suppose that $f(z)$ is mean p-valent and has q zeros (with due count of multiplicity) in a domain Δ. Near a zero of order $h, f(z)$ assumes all sufficiently small values exactly h times. It follows that for small positive R

$$p(R) \geqslant q, \quad W(R) \geqslant qR^2,$$

and since $f(z)$ is mean p-valent, we must have $q \leqslant p$. In particular $f(z)$ can have zeros at all only if $p \geqslant 1$.

We consider then a general function

$$f(z) = a_0 + a_1 z + \dots$$

regular in $|z| < 1$, and write

$$\mu_q = \max_{v \leqslant q} |a_v|.$$

We shall denote by $A(p)$, $A(p,q)$, etc., constants depending on p only or on p and q respectively.

We can then prove the following extension of Theorem 2.2:

THEOREM 2.4. *Suppose that $f(z)$ is regular in $|z| < 1$ and has at most q zeros there, of which not more than h lie in $|z| < \frac{1}{2}$. Set*

$$R_1 = (h+2)\, 2^{h-1}\mu_h, \quad R_2 = M(r,f) \quad (0 < r < 1).$$

Then we have with the notation of (2.1) and (2.3)

$$\int_{R_1}^{R_2} \frac{dR}{Rp(R)} < 2\log\frac{1}{1-r} + A(q).$$

2.4.1. We shall split up the proof of Theorem 2.4, which is a little involved, into a number of stages.

Our first aim is to find a point, near the origin but not too near any zero of $f(z)$, where $|f(z)| \leqslant R_1$.

LEMMA 2.2. *Suppose that* $f(z) = a_0 + a_1 z + \dots$ *is regular in* $|z| \leqslant \rho$, *where* $0 < \rho \leqslant 1$, *and that* $f(z)$ *has at most h zeros there. Then*

$$P = \min_{|z|=\rho} |f(z)| \leqslant (h+2) 2^{h-1} \mu_h.$$

The result is evident if $h = 0$, since in this case $1/f(z)$ is regular in $|z| \leqslant \rho$ and has maximum $1/P$ on $|z| = \rho$. Thus the maximum modulus theorem gives $1/P \geqslant 1/|a_0|$ as required.

Suppose then that $h > 0$ and also $P > 0$, since otherwise there is nothing to prove. We also suppose that $\rho = 1$. For having proved the result in this case we may, if $\rho < 1$, consider $\phi(z) = f(\rho z)$ instead of $f(z)$, and the result for $f(z)$ will follow, since the coefficients of $\phi(z)$ have moduli not larger than those of $f(z)$.

Let $z_1, z_2, \dots, z_h$ be the zeros of $f(z)$ in $|z| \leqslant 1$ and so in $|z| < 1$ and define $g(z)$ by

$$\prod_{n=1}^{h} (z - z_n) = \prod_{n=1}^{h} (1 - \bar{z}_n z) f(z) g(z). \tag{2.9}$$

Then

$$g(z) = \sum_{\nu=0}^{\infty} g_\nu z^\nu$$

is regular in $|z| \leqslant 1$. Also on $|z| = 1$ we have

$$|g(z)| = \left| \frac{1}{f(z)} \right| \leqslant \frac{1}{P}.$$

Hence the Cauchy inequalities yield

$$|g_\nu| \leqslant \frac{1}{P}.$$

Thus the coefficients of the function on the right-hand side of (2.9) are not greater than those of

$$(1+z)^h \sum_{\nu=0}^{\infty} |a_\nu| z^\nu \sum_{\nu=0}^{\infty} P^{-1} z^\nu.$$

Since the coefficient of z^h is unity in the left-hand side of (2.9), we deduce from equating coefficients of z^h in (2.9) that

$$1 \leqslant P^{-1} \sum_{r=0}^{h} \binom{h}{r} \sum_{\nu=0}^{h-r} |a_\nu| \leqslant P^{-1} \sum_{r=0}^{h} (h-r+1) \binom{h}{r} \mu_h$$

$$= (h+2) \, 2^{h-1} P^{-1} \mu_h.$$

This proves the lemma.

2.4.2. Our argument now proceeds by successive applications of Theorem 2.2, moving along a chain of circles so as to avoid the zeros of $f(z)$.

LEMMA 2.3. *Suppose that $f(z)$ is regular in Δ: $|z| < 1$ and has at most q zeros there. Let z_1, z_2 be two points such that the circles $|z - z_1| < \delta$, $|z - z_2| < \delta$ lie in $|z| < 1$ and are free from the zeros of $f(z)$ and that $|z_1 - z_2| < C\delta$, where C, δ are positive. Then if $p(R) = p(R, \Delta)$ is defined as in (2.3) and $R_1 = |f(z_1)|$, $R_2 = |f(z_2)|$, we have*

$$\int_{R_1}^{R_2} \frac{dR}{R p(R)} < A(C, q).$$

In applications C will be a constant depending only on q, so that $A(C, q)$ will also depend only on q.

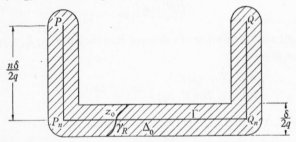

Fig. 2

To prove Lemma 2.3 let P, Q denote the points z_1, z_2. Write $P_0 = P$ and for $1 \leqslant n \leqslant q$ draw PP_n of length $n\delta/(2q)$ and such that the angle $P_n PQ$ is a right angle described in the positive sense. Let $Q = Q_0$ and QQ_n be equal and parallel to PP_n.

Then the lines $P_n Q_n$ for $0 \leqslant n \leqslant q$ are each of length at most $C\delta$, and no point can be distant less than $\delta/(4q)$ from more than one of these lines. Hence at least one of these segments is distant at least $\delta/(4q)$ from all the zeros of $f(z)$. If P_nQ_n is such a segment consider the polygonal arc Γ: PP_nQ_nQ. Since all the corners of Γ lie in $|z| \leqslant 1 - \frac{1}{2}\delta$, so does the whole of Γ. Further, the arc is by construction and hypothesis distant at least $\delta/(4q)$ from every zero of $f(z)$.

Let Δ_0 be the domain consisting of all those points distant at most $\delta/(4q)$ from Γ. The area A of Δ_0 is at most $\dfrac{\delta}{2q}\left[l + \dfrac{\delta}{2q} \right]$, where l is the length of Γ and so

$$A \leqslant (C+2)\frac{\delta^2}{2q}.$$

Also for $R_1 \leqslant R \leqslant R_2$ there is a point z_0 on Γ such that $|f(z_0)| = R$. Since $f(z) \neq 0$ in Δ_0, we see, just as in the proof of Theorem 2.2, that the level curve γ_R: $|f(z)| = R$, though this point z_0 cannot have a loop inside Δ_0, for such a loop would have to contain a zero of $f(z)$. Thus γ_R must stretch to the boundary of Δ_0 in both directions, except for a finite number of values of R, for which γ_R may meet a zero of $f'(z)$.

We may thus apply Theorem 2.1 to $f(z)$ in Δ_0 with

$$l(R, \Delta_0) \geqslant \frac{\delta}{2q}$$

for all but a finite number of values of R in the interval $R_1 \leqslant R \leqslant R_2$, and this gives

$$\left(\frac{\delta}{2q}\right)^2 \int_{R_1}^{R_2} \frac{dR}{Rp(R, \Delta_0)} \leqslant 2\pi A \leqslant 2\pi(C+2)\frac{\delta^2}{4q}.$$

Since $p(R, \Delta) \geqslant p(R, \Delta_0)$, we deduce Lemma 2.3.

2.4.3. We also need

LEMMA 2.4. *Suppose that $f(z)$ is regular in Δ: $|z| < 1$ and has no zeros in the annulus $2r_1 - 1 < |z| < \frac{1}{2}(1 + r_2)$, where $\frac{1}{2} \leqslant r_1 < r_2 < 1$. Write $z_1 = r_1 e^{i\theta}$, $z_2 = r_2 e^{i\theta}$, $|f(z_1)| = R_1$, $|f(z_2)| = R_2$ and define $p(R) = p(R, \Delta, f)$ as in (2.3). Then we have*

$$\int_{R_1}^{R_2} \frac{dR}{Rp(R)} < 2\log\frac{1 - r_1}{1 - r_2} + 10.$$

Put $$z = z_1 + \delta\zeta, \quad \phi(\zeta) = f(z_1 + \delta\zeta),$$

where $\delta = \frac{1}{2}(1 + r_2) - r_1$. Then the circle C_0: $|\zeta| < 1$ corresponds to a circle lying in the annulus $r_1 - \delta < |z| < r_1 + \delta$ and so in $2r_1 - 1 < |z| < \frac{1}{2}(1 + r_2)$, since $\frac{1}{2}(1 + r_2) < 1$. Thus $\phi(\zeta) \neq 0$ in C_0 and

$$p(R, C_0, \phi) \leqslant p(R, \Delta, f) = p(R).$$

If we define ζ by $z_1 + \delta\zeta = z_2$, then Theorem 2.2 applied to $\phi(\zeta)$ gives

$$\int_{R_1}^{R_2} \frac{dR}{Rp(R)} \leqslant 2\left[\log\left(\frac{1 + |\zeta|}{1 - |\zeta|}\right) + \pi\right] < 2\left[\log\left(\frac{2}{1 - |\zeta|}\right) + \pi\right]$$

$$= 2\left[\log\frac{2\delta}{\delta - (r_2 - r_1)} + \pi\right] < 2\left[\log\frac{4(1 - r_1)}{1 - r_2} + \pi\right].$$

This proves the lemma.

2.4.4. Completion of proof of Theorem 2.4.

Suppose now that the hypotheses of Theorem 2.4 hold. Since there are at most $h \leqslant q$ zeros in $|z| < \frac{1}{2}$, at least one of the annuli

$$\nu/[2(q+1)] < |z| < (\nu+1)/[2(q+1)] \quad \text{for} \quad 0 \leqslant \nu \leqslant q,$$

is free from zeros of $f(z)$. If ν is a value for which this is true we choose

$$r_1 = \frac{\nu + \frac{1}{2}}{2(q+1)}.$$

Then by Lemma 2.2 we can find z_1 such that $|z_1| = r_1$ and $|f(z_1)| \leqslant R_1$, where R_1 is defined as in Theorem 2.4. Further the circle $|z - z_1| < [4(q+1)]^{-1}$ lies in $|z| < \frac{1}{2}$ and contains no zero of $f(z)$.

We now surround each zero ζ_μ of $f(z)$ which lies in $\frac{3}{4} \leqslant |z| < 1$ by an annulus $$2|\zeta_\mu| - 1 < |z| < \frac{1}{2}(1 + |\zeta_\mu|).$$

If two or more such annuli overlap, we join them to form a single annulus. We thus obtain at most q non-overlapping annuli, which we shall call 'black' annuli. If $\rho < |z| < \rho'$ is such a black annulus, then $$1 - \rho < 4^q(1 - \rho'). \tag{2.10}$$

We may assume $r \geqslant \frac{3}{4}$ in Theorem 2.4, for if the theorem is true when $r \geqslant \frac{3}{4}$, it is true generally. Let r_2 be the smallest number such that $r_2 \geqslant r$ and $|z| = r_2$ does not lie in a black annulus. Then we have by (2.10)

$$(1 - r_2) \geqslant 4^{-q}(1 - r). \tag{2.11}$$

Also $M(r_2, f) \geqslant M(r, f)$, and so we can find $z_2 = r_2 e^{i\theta}$ such that

$$|f(z_2)| \geqslant R_2 = M(r, f).$$

Let ρ_0' be the lower bound of all numbers such that $\rho_0' \geqslant \frac{3}{4}$ and $|z| = \rho_0'$ does not lie in a black annulus. Then by (2.10)

$$\rho_0' < 1 - 4^{-(q+1)}, \tag{2.12}$$

and also $\frac{3}{4} \leqslant \rho_0' \leqslant r_2$. Let $\rho_1 < |z| < \rho_1', \rho_2 < |z| < \rho_2', \ldots, \rho_k < |z| < \rho_k'$ be the black annuli (if any) separating $|z| = \rho_0'$ and $|z| = r_2$, arranged so that

$$\rho_0' \leqslant \rho_1 < \rho_1' \leqslant \rho_2 < \rho_2' \leqslant \ldots \leqslant \rho_k < \rho_k' \leqslant r_2.$$

Write $\zeta_\nu = \rho_\nu e^{i\theta} \ (1 \leqslant \nu \leqslant k)$, $\zeta_\nu' = \rho_\nu' e^{i\theta} \ (0 \leqslant \nu \leqslant k)$, and write

$$|f(\zeta_\nu)| = R_{(\nu)}, \quad |f(\zeta_\nu')| = R_{(\nu)}'.$$

Then the points ζ_ν, ζ_ν' lie outside black annuli for $1 \leqslant \nu \leqslant k$, and so no zeros of $f(z)$ lie in the circles

$$|z - \zeta_\nu| < \tfrac{1}{2}(1 - \rho_\nu), \quad |z - \zeta_\nu'| < \tfrac{1}{2}(1 - \rho_\nu'),$$

and in view of (2.10) we have

$$|\zeta_\nu' - \zeta_\nu| < 1 - \rho_\nu < 4^q(1 - \rho_\nu').$$

Thus we may apply Lemma 2.3 with $\delta = \tfrac{1}{2}(1 - \rho_\nu')$, $C = 2 \cdot 4^q$ and obtain

$$\int_{R_{(\nu)}}^{R_{(\nu)}'} \frac{dR}{R p(R)} < A(q) < A(q) + 2 \log \left(\frac{1 - \rho_\nu}{1 - \rho_\nu'} \right) \quad (1 \leqslant \nu \leqslant k). \tag{2.13}$$

Similarly the circles

$$|z - z_1| < \frac{1}{4(q+1)}, \quad |z - \zeta_0'| < \tfrac{1}{2}(1 - \rho_0')$$

contain no zeros, and in view of (2.12) we may apply Lemma 2.3 which yields

$$\int_{R_1}^{R_{(0)}} \frac{dR}{R p(R)} < A(q). \tag{2.14}$$

Finally, no zeros lie in the annuli

$$2\rho'_\nu - 1 < |z| < \tfrac{1}{2}(1 + \rho_{\nu+1}) \quad (0 \leqslant \nu \leqslant k-1),$$

or
$$2\rho'_k - 1 < |z| < \tfrac{1}{2}(1 + r_2),$$

and so we may apply Lemma 2.4 and obtain

$$\int_{R'_{(\nu)}}^{R_{(\nu+1)}} \frac{dR}{Rp(R)} < 2\log\frac{1-\rho'_\nu}{1-\rho_{\nu+1}} + 10 \quad (0 \leqslant \nu \leqslant k-1), \quad (2.15)$$

and also

$$\int_{R'_{(k)}}^{R_2} \frac{dR}{Rp(R)} < 2\log\frac{1-r_2}{1-\rho'_k} + 10. \quad (2.16)$$

On adding the inequalities (2.13) to (2.16) we obtain

$$\int_{R_1}^{R_2} \frac{dR}{Rp(R)} < 2\log\frac{1-\rho'_0}{1-r_2} + [10 + A(q)](k+1)$$

$$< 2\log\frac{1}{1-r} + A(q),$$

in view of (2.11) and since $k \leqslant q$. This proves Theorem 2.4.

2.5. The theorem of Cartwright and Spencer. We can now eliminate the hypothesis $f(z) \neq 0$ from Theorem 2.3 also and prove

THEOREM 2.5. *Suppose that* $f(z) = \sum_{0}^{\infty} a_n z^n$ *is mean* p-*valent in* $|z| < 1$. *Then*

$$M(r, f) < A(p)\mu_p(1-r)^{-2p} \quad (0 < r < 1).$$

The result was proved by Cartwright [1] for p-valent functions. In its present form it is due to Spencer [3].

We note that since $f(z)$ is mean p-valent in $|z| < 1$, $f(z)$ has $q \leqslant p$ zeros there. Now Theorem 2.4 gives

$$\int_{R_1}^{R_2} \frac{dR}{Rp(R)} < 2\log\frac{1}{1-r} + A(q),$$

and since $f(z)$ is mean p-valent in $|z| < 1$, so that $H(R) \leqslant 0$ in Lemma 2.1, we deduce from that lemma

$$\int_{R_1}^{R_2} \frac{dR}{Rp(R)} \geqslant \frac{1}{p}\log\frac{R_2}{R_1} - \frac{1}{2}.$$

Thus we obtain

$$\log\frac{R_2}{R_1} < 2p\log\frac{1}{1-r} + \frac{p}{2} + pA(q),$$

and bearing in mind the definitions of R_1, R_2 in Theorem 2.4 and the inequalities $0 \leqslant h \leqslant q \leqslant p$, we have

$$M(r, f) < A(p)\mu_h(1-r)^{-2p}$$

$$\leqslant A(p)\mu_p(1-r)^{-2p},$$

as required.

2.6. Simultaneous growth near different boundary points.†

We have seen that a function $f(z)$ mean p-valent in $|z| < 1$ satisfies

$$|f(z)| = O(1-r)^{-2p} \quad (|z| = r).$$

However, a function can be as large as this only on a single rather small arc of $|z| = r$. We shall close this chapter by investigating in what way $f(z)$ can become large near several distant points of $|z| = r$. Our basic result is

THEOREM 2.6. *Suppose that $f(z)$ is mean p-valent in a domain Δ containing k non-overlapping circles $|z - z_n| < r_n$ $(1 \leqslant n \leqslant k)$. Suppose further that $|f(z_n)| \leqslant R_1$, $|f(z_n')| \geqslant R_2 > eR_1$, where*

$$\delta_n = \frac{r_n - |z_n' - z_n|}{r_n} > 0,$$

and that $f(z) \neq 0$ for $|z - z_n| < \frac{1}{2}r_n$ $(1 \leqslant n \leqslant k)$. Then

$$\sum_{n=1}^{k}\left[\log\frac{A(p)}{\delta_n}\right]^{-1} < \frac{2p}{\log(R_2/R_1) - 1}.$$

Let Δ_n be the circle $|z - z_n| < r_n$ and let

$$p_n(R) = p(R, \Delta_n, f)$$

be defined as in (2.3). Consider

$$\phi(\zeta) = f(z_n + r_n\zeta).$$

† The reader may, if he wishes, defer the remaining results of this chapter until they are required in Chapters 3 and 5.

Then $p_n(R)$ corresponds to $\phi(\zeta)$ and $|\zeta| < 1$. We choose ζ_n so that

$$z_n + r_n \zeta_n = z'_n, \quad \zeta_n = \frac{z'_n - z_n}{r_n}.$$

Then $|\phi(0)| \leqslant R_1$, $|\phi(\zeta_n)| \geqslant R_2$, $\phi(\zeta) \neq 0$ for $|\zeta| < \frac{1}{2}$ and $\phi(\zeta)$ has at most p zeros in $|\zeta| < 1$. Thus Theorem 2.4 gives

$$\int_{R_1}^{R_2} \frac{dR}{R p_n(R)} < 2 \log \frac{A(p)}{1 - |\zeta_n|} = 2 \log \frac{A(p)}{\delta_n}.$$

Now we have from Schwarz's inequality

$$\left(\log \frac{R_2}{R_1} \right)^2 = \left(\int_{R_1}^{R_2} \frac{dR}{R} \right)^2 \leqslant \int_{R_1}^{R_2} \frac{p_n(R)\,dR}{R} \int_{R_1}^{R_2} \frac{dR}{R p_n(R)},$$

and so

$$\frac{1}{\log \left(\dfrac{A(p)}{\delta_n} \right)} \leqslant \frac{2}{\displaystyle\int_{R_1}^{R_2} \frac{dR}{R p_n(R)}} \leqslant \frac{2}{\left(\log \dfrac{R_2}{R_1} \right)^2} \int_{R_1}^{R_2} \frac{p_n(R)\,dR}{R}.$$

Adding, we deduce

$$\sum_{n=1}^{k} \left[\log \frac{A(p)}{\delta_n} \right]^{-1} \leqslant \frac{2}{[\log (R_2/R_1)]^2} \int_{R_1}^{R_2} \left[\sum_{n=1}^{k} p_n(R) \right] \frac{dR}{R}. \quad (2.17)$$

Now

$$\sum_{n=1}^{k} p_n(R) = \sum_{n=1}^{k} p(R, \Delta_n) \leqslant p(R, \Delta) = p(R). \quad (2.18)$$

Using the notation of §2.3 we have, since $f(z)$ is mean p-valent in Δ,

$$\int_{R_1}^{R_2} p(R) \frac{dR}{R} = p \log \frac{R_2}{R_1} + \int_{R_1}^{R_2} h(R) \frac{dR}{R}$$

$$= p \log \frac{R_2}{R_1} + \frac{1}{2} \int_{R_1}^{R_2} \frac{dH(R)}{R^2}$$

$$= p \log \frac{R_2}{R_1} + \frac{H(R_2)}{2R_2^2} - \frac{H(R_1)}{2R_1^2} + \int_{R_1}^{R_2} \frac{H(R)\,dR}{R^3}$$

$$\leqslant p \left[\log \left(\frac{R_2}{R_1} \right) + \frac{1}{2} \right],$$

since $-pR^2 \leqslant H(R) \leqslant 0$. Using (2.17) and (2.18), we deduce

$$\sum_{n=1}^{k} \left[\log\left(\frac{A(p)}{\delta_n}\right) \right]^{-1} \leqslant \frac{2p}{[\log(R_2/R_1)]} + \frac{p}{[\log(R_2/R_1)]^2}$$
$$< \frac{2p}{\log(R_2/R_1)-1},$$

and this proves Theorem 2.6.

2.7. We shall make two applications of Theorem 2.6. We first apply it to the concept of order of a mean p-valent function at a boundary point.

Suppose then that $f(z)$ is mean p-valent in $|z| < 1$ and that for $\zeta = e^{i\theta}$ there is a path $\gamma(\theta)$, lying except for its end-point ζ in $|z| < 1$, and also a positive δ such that

$$\varlimsup (1-|z|)^\delta |f(z)| > 0,$$

as $z \to \zeta$ along $\gamma(\theta)$. Then we define the *order* $\alpha(\zeta)$ of $f(z)$ at ζ as the upper bound of all such δ. If no path $\gamma(\theta)$ and positive δ exist, we put $\alpha(\zeta) = 0$. We can then prove the following result of Spencer [3].

THEOREM 2.7. *If $f(z)$ is mean p-valent in $|z| < 1$, then the set E of distinct ζ on $|\zeta| = 1$ such that $\alpha(\zeta) > 0$ is countable and satisfies $\sum_E \alpha(\zeta) \leqslant 2p$.*

It suffices to show that if $\zeta_1, \zeta_2, ..., \zeta_k$ are distinct points of $|\zeta| = 1$, then

$$\sum_{n=1}^{k} \alpha(\zeta_n) \leqslant 2p.$$

For then the set E_N of ζ, such that $|\zeta| = 1$ and $\alpha(\zeta) > N^{-1}$, is finite for $N = 1, 2, 3, ...$, and so the set E consisting of all the E_N is countable. Letting $k \to \infty$ in the above inequality we have $\Sigma \alpha(\zeta) \leqslant 2p$ as required.

Suppose then that Theorem 2.7 is false. Then by the above remark we can find a finite set of points $\zeta_1, \zeta_2, ..., \zeta_k$ and $\epsilon > 0$ such that

$$\sum_{n=1}^{k} \alpha(\zeta_n) = 2(p + k\epsilon).$$

For each ζ_n there exists a path γ_n, approaching ζ_n from $|z| < 1$, on which

$$(1-|z|)^{\eta_n} |f(z)| > 1,$$

where $\eta_n = \alpha(\zeta_n) - \epsilon$, and so

$$\sum_{n=1}^{k} \eta_n > 2p. \qquad (2.19)$$

Hence there exists $R_0 > 0$, such that for $R_2 > R_0$ we can find $z'_n = r_n e^{i\theta_n}$ on γ_n, such that

$$|f(z'_n)| = R_2 > \left(\frac{1}{1-r_n}\right)^{\eta_n} \quad (1 \leqslant n \leqslant k). \qquad (2.20)$$

Choose now δ so small that

(i) $f(z)$ has no zeros for $1 - 2\delta < |z| < 1$,

(ii) $4\delta < \min_{1 \leqslant m < n \leqslant k} |\zeta_m - \zeta_n|$,

and set $r_0 = 1 - \delta$. Then if R_2 is sufficiently large, we have

$$|z'_m - z'_n| > 4\delta \quad (1 \leqslant m < n \leqslant k),$$

since $z'_n \to \zeta_n$ as $R_2 \to \infty$, and if we put $z_n = r_0 e^{i\theta_n}$ it follows that the circles $|z - z_n| < \delta$ are mutually exclusive. Also

$$|f(z_n)| \leqslant R_1 = M(r_0, f) \quad \text{and} \quad |f(z'_n)| = R_2,$$

where we may suppose $R_2 > eR_1$. Thus we may apply Theorem 2.6 with

$$\delta_n = \frac{\delta - (r_n - r_0)}{\delta} = \frac{1 - r_n}{\delta},$$

and obtain

$$\sum_{n=1}^{k} \left[\log\left(\frac{A(p)\delta}{1-r_n}\right)\right]^{-1} \leqslant \frac{2p}{\log(R_2/eR_1)}.$$

In view of (2.20) this gives

$$\sum_{n=1}^{k} [\log(A(p)\delta R_2^{1/\eta_n})]^{-1} \leqslant \frac{2p}{\log R_2 - \log(eR_1)},$$

$$\sum_{n=1}^{k} \frac{\eta_n}{\eta_n \log[A(p)\delta] + \log R_2} \leqslant \frac{2p}{\log R_2 - \log(eR_1)}.$$

As $R_2 \to \infty$ the two sides of this inequality are asymptotically equal to

$$\frac{\Sigma \eta_n}{\log R_2} \quad \text{and} \quad \frac{2p}{\log R_2}$$

respectively, and so we obtain a contradiction from (2.19). This proves Theorem 2.7.

2.7.1. An example. Theorem 2.7 is best possible. Suppose in fact that α_n is any sequence of positive numbers for $1 \leqslant n < k$, where k may be finite or infinite, such that

$$\Sigma \alpha_n = 2.$$

Set $S_0 = 0$, $S_N = \sum_{n=1}^{N} \alpha_n$ $(1 \leqslant N < k)$, and let D be the domain consisting of the w plane cut from $|w| = 1$ to ∞ along the lines

$$\arg w = \pi S_N \quad (0 \leqslant N < k).$$

By Riemann's mapping theorem† we can find a function $z = \phi(w)$ which maps $D\,(1, 1)$ conformally onto $|z| < 1$.

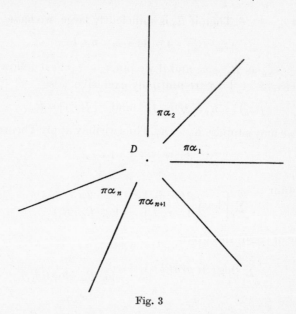

Fig. 3

As w moves along the boundary of that part D_N of D which lies in the angle $\pi S_{N-1} < \arg w < \pi S_N$, z moves along an arc of the unit circle. Simultaneously $W = w^{-1/\alpha_n}$ moves along a straight line segment through the origin and so by Schwarz's reflexion principle‡ z can be continued as a regular function of W near

† See, for example, Ahlfors[2], p. 172.
‡ See, for example, Ahlfors[2], p. 193.

$W = 0$, and conversely if $z = z_n$ corresponds to $W = 0$, W becomes a regular function of z near z_n. Thus

$$W = c_1(z - z_n) + c_2(z - z_n)^2 + \dots, \quad \text{near } z = z_n,$$

where $c_1 \neq 0$. This gives

$$w \sim [c_1(z - z_n)]^{-\alpha_n} \quad (z \to z_n).$$

Hence $w = f(z) = \phi^{-1}(z)$, which maps $|z| < 1$ onto D, is a univalent function having the orders α_n at the points z_n for $1 \leqslant n < k$, and the α_n are arbitrary subject to $\Sigma \alpha_n = 2$. Again for any positive p the function $[f(z) - 1]^p$ is mean p-valent in $|z| < 1$ and has orders $p\alpha_n$ at the points z_n, where $p\alpha_n$ is now an arbitrary sequence satisfying $\Sigma p\alpha_n = 2p$.

2.8. Functions of maximal growth.

We have seen that if $f(z)$ is mean p-valent in $|z| < 1$ then

$$M(r, f) = O(1 - r)^{-2p} \quad (r \to 1).$$

We complete the chapter by investigating more closely those functions for which this order of growth is effectively attained so that

$$\alpha = \overline{\lim_{r \to 1}} (1 - r)^{2p} M(r, f) > 0. \tag{2.21}$$

We show that $f(z)$ attains this growth along a certain radius and that $|f(z)|$ is quite small except near this radius.

THEOREM 2.8. *Suppose that $f(z)$ is mean p-valent in $|z| < 1$ and that (2.21) holds. Then there exists θ_0 $(0 \leqslant \theta_0 < 2\pi)$ such that*

$$\alpha_0 = \lim_{r \to 1} (1 - r)^{2p} |f(r\, e^{i\theta_0})| \geqslant \frac{\alpha}{A(p)}.$$

Since $f(z)$ can have only a finite number of zeros in $|z| < 1$ we may suppose that $f(z) \neq 0$ in $1 - 2\delta < |z| < 1$, where $\delta > 0$. We set $r_0 = 1 - \delta$. Then there exists a sequence $\zeta_n = r_n e^{i\theta_n}$, where

$$r_0 < r_n < 1, \quad 0 \leqslant \theta_n \leqslant 2\pi,$$

such that $r_n \to 1$ $(n \to \infty)$ and

$$|f(\zeta_n)| > \tfrac{1}{2}\alpha(1 - r_n)^{-2p}.$$

We take $z_1 = r\,e^{i\theta_n}$, where $r_0 < r < r_n$ and $z_2 = r_n\,e^{i\theta_n} = \zeta_n$, put $R_1 = |f(z_1)|$, $R_2 = |f(z_2)|$ and apply Lemma 2.4. This gives

$$\int_{R_1}^{R_2} \frac{dR}{Rp(R)} < 2\log\frac{1-r}{1-r_n} + 10.$$

Since $f(z)$ is mean p-valent we have from Lemma 2.1

$$\int_{R_1}^{R_2} \frac{dR}{Rp(R)} \geqslant \frac{1}{p}\left\{\log\frac{R_2}{R_1} - \frac{1}{2}\right\}.$$

Hence

$$|f(r\,e^{i\theta_n})|\,(1-r)^{2p} = R_1(1-r)^{2p} > \frac{R_2(1-r_n)^{2p}}{A(p)} > \frac{\alpha}{A(p)} \quad (r_0 < r < r_n).$$

Let θ_0 be a limit point of the sequence θ_n. We deduce

$$|f(r\,e^{i\theta_0})|\,(1-r)^{2p} \geqslant \frac{\alpha}{A(p)} \quad (r_0 \leqslant r < 1),$$

since $r_n \to 1$ $(n \to \infty)$, and this proves Theorem 2.8.

We shall be able to show in Chapter 5 that under the stronger assumption (2.8) $\alpha = \alpha_0$, so that both quantities exist as limits.

2.8.1. It follows from Theorem 2.7 that $f(z)$ must have zero order at all points of $|z| = 1$ other than $e^{i\theta_0}$, so that $e^{i\theta_0}$ is necessarily unique in Theorem 2.8. However, we can prove more than this.

THEOREM 2.9. *With the hypothesis of Theorem 2.8 and given ϵ such that $0 < \epsilon < 2p$, we can find a positive constant C and $r_0 < 1$ such that*

$$|f(r\,e^{i\theta})| < \frac{1}{(1-r)^\epsilon\,|\theta-\theta_0|^{2p-\epsilon}} \tag{2.22}$$

for $r_0 < r < 1$, $C(1-r) \leqslant |\theta-\theta_0| \leqslant \pi$. Further we have uniformly as $r \to 1$, while $\epsilon \leqslant |\theta-\theta_0| \leqslant \pi$,

$$\log|f(r\,e^{i\theta})| < O\left[\log\frac{1}{1-r}\right]^{\frac{1}{2}}. \tag{2.23}$$

In particular, we see that (2.23) holds as $r\,e^{i\theta}$ approaches any point on $|z| = 1$ distinct from $e^{i\theta_0}$. This is a good deal stronger than the condition for zero order.

Suppose then that $f(z)$ is mean p-valent, $f(z) \neq 0$ in

$$1 - 2\delta < |z| < 1,$$

and that $\qquad |f(r e^{i\theta_0})| \geqslant \tfrac{1}{2}\alpha_0(1-r)^{-2p} \quad (1 - \delta < r < 1).$ $\qquad$ (2.24)

These assumptions are satisfied for all sufficiently small δ. We also write

$$R_1 = M(1-\delta, f) = \max_{|z| = 1-\delta} |f(z)|,$$

$$z_1 = (1-\delta)\, e^{i\theta_0}, \quad z_2 = (1-\delta)\, e^{i\theta},$$

and assume that $4\delta \leqslant |\theta - \theta_0| \leqslant \pi$. Then the circles $|z - z_1| < \delta$, $|z - z_2| < \delta$ are mutually exclusive.

We now suppose further that

$$|f(r_2 e^{i\theta})| = R_2 > eR_1, \qquad (2.25)$$

where $1 - \delta < r_2 < 1$, and put $z_2' = r_2 e^{i\theta}$, $z_1' = r_1 e^{i\theta_0}$, where r_1 is chosen to be the smallest number such that

$$|f(z_1')| = |f(r_1 e^{i\theta_0})| = R_2.$$

Such a number exists by Theorem 2.8 and (2.25). Then $r_1 > 1 - \delta$ by definition of R_1.

We now apply Theorem 2.6 with

$$\delta_1 = \frac{\delta - [r_1 - (1-\delta)]}{\delta} = \frac{1 - r_1}{\delta}, \quad \delta_2 = \frac{1 - r_2}{\delta},$$

and obtain

$$\left[\log\left(\frac{A(p)\,\delta}{1 - r_2}\right)\right]^{-1} \leqslant \frac{2p}{\log(R_2/R_1) - 1} - \left[\log\left(\frac{A(p)\,\delta}{1 - r_1}\right)\right]^{-1}. \quad (2.26)$$

Now by (2.24) $\qquad R_2 \geqslant \tfrac{1}{2}\alpha_0(1 - r_1)^{-2p},$

and by Theorem 2.5

$$R_1 \leqslant A(p)\,\mu_p\,\delta^{-2p}.$$

Thus $\qquad\qquad \dfrac{R_2}{R_1} \geqslant C_1\left(\dfrac{\delta}{1 - r_1}\right)^{2p},$

where $C_1, C_2, \ldots,$ will denote constants depending on $f(z)$ and p only.

We deduce from this and (2.25) that

$$\left[\log\frac{A(p)\,\delta}{1-r_2}\right]^{-1} \leqslant \frac{2p}{\log\left(R_2/R_1\right)-1} - \frac{2p}{\log\left(R_2/R_1\right)+C_2},$$

and this gives

$$\log\left(\frac{A(p)\,\delta}{1-r_2}\right) \geqslant C_3\left\{\log\left(\frac{R_2}{R_1}\right)-1\right\}^2.$$

Thus

$$\log\frac{R_2}{R_1} \leqslant 1 + C_4\left[\log\left(\frac{C_5\delta}{1-r_2}\right)\right]^{\frac{1}{2}}. \tag{2.27}$$

Taking δ and R_1 fixed, we deduce (2.23), provided that (2.25) holds, and the inequality is trivial otherwise.

It remains to prove (2.22). We may again without loss in generality suppose that (2.25) holds. Then (2.27) gives, if $\delta = \frac{1}{4}\,|\,\theta-\theta_0\,| \geqslant 1-r_2$,

$$R_2 < R_1 \exp\left\{1 + C_4\left[\log\frac{C_5\delta}{(1-r_2)}\right]^{\frac{1}{2}}\right\}$$

$$< C_6\,|\,\theta-\theta_0\,|^{-2p} \exp\left\{C_4\left[\log\frac{C_5\delta}{1-r_2}\right]^{\frac{1}{2}}\right\}$$

$$< \frac{1}{|\,\theta-\theta_0\,|^{2p-\epsilon}\,(1-r_2)^{\epsilon}},$$

provided that

$$\left(\frac{|\,\theta-\theta_0\,|}{1-r_2}\right)^{\epsilon} > C_6 \exp\left\{C_4\left(\log\left[\frac{C_5\,|\,\theta-\theta_0\,|}{4(1-r_2)}\right]\right)^{\frac{1}{2}}\right\},$$

which is true for $|\,\theta-\theta_0\,| \geqslant C(1-r_2)$, where C depends on p, ϵ and $f(z)$ only. This proves (2.22) and completes the proof of Theorem 2.9.

Some further applications of Theorem 2.6 will be given in later chapters.

CHAPTER 3

MEANS AND COEFFICIENTS

3.0. Introduction. In the last chapter we investigated the growth of a function

$$f(z) = \sum_0^\infty a_n z^n,$$

mean p-valent in $|z| < 1$. We showed in Theorem 2.5 that the maximum modulus $M(r, f)$ satisfies

$$M(r, f) < A(p)\mu_p(1-r)^{-2p} \quad (0 < r < 1). \tag{3.1}$$

In this chapter we estimate the order of magnitude of the coefficients a_n and show that, for any function $f(z)$ mean p-valent in $|z| < 1$ and constants $C > 0$ and $\alpha > \frac{1}{2}$, the inequality

$$M(r, f) < C(1-r)^{-\alpha} \quad (0 < r < 1), \tag{3.2}$$

implies $\quad |a_n| < A(p, \alpha) C(1+n)^{\alpha-1} \quad (n = 0, 1, 2, \ldots). \tag{3.3}$

It will follow at once that, if $f(z)$ is mean p-valent in $|z| < 1$, so that (3.1) holds, then

$$|a_n| < A(p)\mu_p n^{2p-1} \quad (n \geqslant 1), \tag{3.4}$$

provided that $p > \frac{1}{4}$. The functions

$$f(z) = (1-z)^{-2p} = \sum_0^\infty b_{n,p} z^n, \tag{3.5}$$

which are mean p-valent in $|z| < 1$ and for which

$$b_{n,p} = \frac{\Gamma(n+2p)}{\Gamma(2p)\,\Gamma(n+1)} \sim \frac{n^{2p-1}}{\Gamma(2p)} \quad (n \to \infty), \tag{3.6}$$

show that the order of magnitude of the bounds in (3.4) is correct.

The method used by Littlewood[1] for proving Theorem 1.6 is sufficient to show that (3.2) implies (3.3) if $f(z)$ is univalent and $\alpha > 1$. The idea for extending this to the case $\alpha > \frac{1}{2}$ occurs first in a joint paper of Littlewood and Paley [1]. The argument was extended to p-valent functions by Biernacki[1] and to mean p-valent functions by Spencer[2].

We shall show further, by means of an example of Spencer[3], that (3.2) does not imply (3.3) for a general mean p-valent function $f(z)$ if $\alpha < \frac{1}{2}$ and that (3.4) is false in general if $p < \frac{1}{4}$.

In the final sections of the chapter we shall give some further applications of our main results by estimating the coefficients of certain classes of mean p-valent functions for which more restrictive bounds than (3.1) can be obtained.

3.1. The Hardy-Stein-Spencer identities.

We suppose now that $f(z)$ is regular in $|z| < 1$ and further that $\lambda > 0$ and $0 < r < 1$. Let $n(r, w)$ be the number of roots of the equation $f(z) = w$ in $|z| < r$ and write

$$p(r, R) = \frac{1}{2\pi} \int_0^{2\pi} n(r, R\, e^{i\psi})\, d\psi.$$

Thus $p(R) = p(r, R)$ is defined as in (2.3) when Δ is the domain $|z| < r$. We also write

$$I_\lambda(r, f) = \frac{1}{2\pi} \int_0^{2\pi} |f(r\, e^{i\theta})|^\lambda\, d\theta.$$

We then have the following remarkable triple identity:†

THEOREM 3.1. *With the above notation*

$$r \frac{d}{dr} I_\lambda(r) = \frac{\lambda^2}{2\pi} \int_0^r \rho\, d\rho \int_0^{2\pi} |f(\rho\, e^{i\theta})|^{\lambda-2} |f'(\rho\, e^{i\theta})|^2\, d\theta$$

$$= \lambda^2 \int_0^\infty p(r, R)\, R^{\lambda-1}\, dR.$$

Suppose first that $f(z)$ has no zero on $|z| = r$ and write

$$f(r\, e^{i\theta}) = R\, e^{i\Phi}.$$

Then near a fixed point of $|z| = r$, we have

$$\log f = \log R + i\Phi, \quad \frac{1}{r} \frac{\partial \Phi}{\partial \theta} = \frac{1}{R} \frac{\partial R}{\partial r},$$

† The first equality is due to Hardy[1] and Stein[1] and the second to Spencer[1].

by the Cauchy-Riemann equations. Thus

$$r\frac{d}{dr}\int_{-\pi}^{+\pi}|f(re^{i\theta})|^{\lambda}d\theta = \lambda\int_{|z|=r}R^{\lambda-1}r\frac{\partial R}{\partial r}d\theta = \lambda\int_{|z|=r}R^{\lambda}d\Phi. \quad (3.7)$$

We make a transformation (not conformal) by writing, when $w = Re^{i\Phi}$, $\qquad P = R^{\frac{1}{2}\lambda}, \quad \psi = \Phi, \quad W = Pe^{i\psi}.$

Then the right-hand side of (3.7) reduces to

$$\lambda\int_{|z|=r}P^2 d\Phi.$$

Now $\frac{1}{2}P^2 d\Phi$ is a sectorial element of area in the W plane, and so the right-hand side of (3.7) represents 2λ times the area in the W plane corresponding to $|z| < r$, multiple points being counted multiply. This is quite evident if $f(z)$ is univalent in $|z| \leqslant r$, so that the area is the interior of the simple closed Jordan curve which is the image of $|z| = r$. In the general case we can prove our result by splitting the disk into a finite number of regions in each of which $f(z)$ is univalent and noting that

$$\int P^2 d\Phi$$

taken over the boundary of the region is additive and so is the area in the W plane.

Now the area in the W plane is equal to

$$\int_0^\infty \int_0^{2\pi} \nu(Pe^{i\psi})P\,dP\,d\psi,$$

where $\nu(Pe^{i\psi})$ is the number of points in $|z| < r$, corresponding to $W = Pe^{i\psi}$. Thus $\quad \nu(Pe^{i\psi}) = n(r, P^{2/\lambda}e^{i\psi}),$

and we obtain

$$\lambda\int_{|z|=r}R^{\lambda}d\Phi = \lambda\int_0^\infty d(P^2)\int_0^{2\pi}\nu(Pe^{i\psi})\,d\psi$$

$$= \lambda\int_{R=0}^\infty (dR^{\lambda})\int_0^{2\pi}n(r, Re^{i\psi})\,d\psi$$

$$= 2\pi\lambda^2\int_0^\infty p(r, R)R^{\lambda-1}dR,$$

and on combining this with (3.7) we see that the first term in the identity of Theorem 3.1 is equal to the third.

Again $|f'(\rho e^{i\theta})|^2 \rho\, d\rho\, d\theta$ is the area of the image of a small element of area, $\rho < |z| < \rho + d\rho$, $\theta < \arg z < \theta + d\theta$ by $w = f(z)$ and so

$$\int_0^r \rho\, d\rho \int_0^{2\pi} |f'(\rho e^{i\theta})|^2 |f(\rho e^{i\theta})|^{\lambda-2} d\theta = \int_0^\infty R\, dR \int_0^{2\pi} R^{\lambda-2} n(r, R e^{i\psi})\, d\psi.$$

In fact both sides represent the total mass in the w plane if a mass density $|w|^{\lambda-2}$ is spread over the image of $|z| < r$ by $w = f(z)$. The right-hand side becomes

$$2\pi \int_0^\infty p(r, R)\, R^{\lambda-1} dR,$$

and so we have the identity of the second and third terms in Theorem 3.1, and that theorem is proved on the assumption that $f(z)$ has no zeros on $|z| = r$.

The result follows in the general case from considerations of continuity. In fact the continuous function $I_\lambda(r)$ has a continuous derivative

$$\frac{S_\lambda(r)}{r} = \frac{\lambda^2}{2\pi r} \int_0^r \rho\, d\rho \int_0^{2\pi} |f(\rho e^{i\theta})|^{\lambda-2} |f'(\rho e^{i\theta})|^2 d\theta,$$

except possibly at certain isolated values of r. At these latter values r_0, $I_\lambda(r)$ clearly remains continuous and so we see that the equation

$$r \frac{d}{dr} I_\lambda(r) = S_\lambda(r)$$

continues to hold, by using the strong form of the mean-value theorem
$$I_\lambda(r_1) - I_\lambda(r_0) = (\log r_1 - \log r_0)\, S_\lambda(\rho),$$
and making r_1 tend to r_0 from below or above.

3.2. Estimates of the means $I_\lambda(r)$.† Suppose again that

$$f(z) = \sum_0^\infty a_n z^n$$

is regular in $|z| < 1$. Then we have for $0 < r < 1$

$$n\,|a_n| = \left| \frac{1}{2\pi i} \int_{|z|=r} \frac{f'(z)\, dz}{z^n} \right| \leqslant \frac{I_1(r, f')}{r^{n-1}}.$$

---

† The results from here to §3.5 inclusive are mainly due to Spencer[2].

We choose $r = n/(n+1)$ for $n \geqslant 1$, so that

$$r^{-(n-1)} = \left(1 + \frac{1}{n}\right)^{n-1} < e,$$

and deduce $\quad |a_n| < \dfrac{e}{n} I_1\left(\dfrac{n}{n+1}, f'\right) \quad (n \geqslant 1).$ $\qquad$ (3.8)

It is thus important to be able to estimate $I_1(r, f')$. For this purpose we use Theorem 3.1. It follows from this theorem that $I_\lambda(r, f)$ is an increasing convex function of $\log r$, when $\lambda > 0$. We have further

THEOREM 3.2. *Suppose that $f(z)$ is mean p-valent in $|z| < 1$ and set $\Lambda = \max(\lambda, \frac{1}{2}\lambda^2)$, when $\lambda > 0$. Then*

$$S_\lambda(r, f) = r \frac{d}{dr} I_\lambda(r, f) \leqslant p\Lambda M(r, f)^\lambda \quad (0 < r < 1), \qquad (3.9)$$

and

$$I_\lambda(r, f) \leqslant M(r_0, f)^\lambda + p\Lambda \int_{r_0}^r \frac{M(t, f)^\lambda \, dt}{t} \quad (0 < r_0 < r < 1). \quad (3.10)$$

We have by Theorem 3.1

$$S_\lambda(r) = \lambda^2 \int_0^\infty p(r, R) R^{\lambda-1} \, dR = \lambda^2 \int_0^{M(r,f)} p(r, R) R^{\lambda-1} \, dR.$$

Also since $f(z)$ is mean p-valent in $|z| < 1$ and so *a fortiori* in $|z| < r$, we have, using the notation of § 2.3,

$$W(R) = \int_0^R p(r, \rho) \, d(\rho^2) \leqslant pR^2 \quad (0 < R < \infty).$$

Hence, writing $M = M(r, f)$, we obtain

$$\int_0^M p(r, R) R^{\lambda-1} \, dR = \frac{1}{2} \int_0^M R^{\lambda-2} \, dW(R)$$

$$= \tfrac{1}{2} M^{\lambda-2} W(M) - \frac{\lambda-2}{2} \int_0^M R^{\lambda-3} W(R) \, dR.$$

There are now two cases. If $\lambda > 2$, we deduce, since $W(R) \geqslant 0$,

$$\int_0^M p(r, R) R^{\lambda-1} \, dR \leqslant \tfrac{1}{2} M^{\lambda-2} W(M) \leqslant \frac{p}{2} M^\lambda = \frac{p}{2} M(r, f)^\lambda.$$

If $0 < \lambda \leqslant 2$ we deduce

$$\int_0^M p(r, R) R^{\lambda-1} dR \leqslant \tfrac{1}{2} M^{\lambda-2} p M^2 + \frac{2-\lambda}{2} \int_0^M R^{\lambda-3} p R^2 dR$$

$$= \frac{p}{2} M^\lambda + \frac{p(2-\lambda)}{2\lambda} M^\lambda = \frac{p}{\lambda} M(r, f)^\lambda.$$

This gives (3.9). Also

$$I_\lambda(r, f) = I_\lambda(r_0, f) + \int_{r_0}^r S_\lambda(t, f) \frac{dt}{t} \leqslant M(r_0, f)^\lambda + p\Lambda \int_{r_0}^r M(t, f)^\lambda \frac{dt}{t},$$

and this yields (3.10).

3.3. Estimates of the coefficients. We now prove our basic result.

THEOREM 3.3. *Suppose that* $f(z) = \sum_0^\infty a_n z^n$ *is mean p-valent in* $|z| < 1$ *and that*

$$M(r, f) \leqslant C(1-r)^{-\alpha} \quad (0 < r < 1), \tag{3.11}$$

where $C > 0$ *and* $\alpha > \tfrac{1}{2}$. *Then we have*

$$|a_n| \leqslant A(p, \alpha) C n^{\alpha-1} \quad (n \geqslant 1). \tag{3.12}$$

We shall need the following preliminary result:

LEMMA 3.1. *Suppose that* $f(z)$ *is mean p-valent in* $|z| < 1$ *and that* $\tfrac{1}{2} \leqslant r < 1$, $0 < \lambda \leqslant 2$. *Then there exists* ρ *such that* $2r - 1 \leqslant \rho \leqslant r$ *and*

$$\frac{1}{2\pi} \int_0^{2\pi} |f'(\rho e^{i\theta})|^2 |f(\rho e^{i\theta})|^{\lambda-2} d\theta \leqslant \frac{4p M(r, f)^\lambda}{\lambda(1-r)}. \tag{3.13}$$

We deduce from Theorems 3.1 and 3.2 that

$$\frac{1}{2\pi} \int_{2r-1}^r \rho \, d\rho \int_0^{2\pi} |f'(\rho e^{i\theta})|^2 |f(\rho e^{i\theta})|^{\lambda-2} d\theta \leqslant \frac{1}{\lambda^2} S_\lambda(r) \leqslant \frac{p}{\lambda} M(r, f)^\lambda.$$

Hence we can choose ρ so that $2r - 1 < \rho < r$ and

$$\frac{1}{2\pi} \int_0^{2\pi} |f'(\rho e^{i\theta})|^2 |f(\rho e^{i\theta})|^{\lambda-2} d\theta \leqslant \frac{p M(r, f)^\lambda}{\tfrac{1}{2}\lambda[r^2 - (2r-1)^2]} \leqslant \frac{4p M(r, f)^\lambda}{\lambda(1-r)}.$$

This proves Lemma 3.1.

We now suppose $r \geqslant \frac{1}{2}$ and that (3.11) holds with $C = 1$ and $\alpha > \frac{1}{2}$. We set $\lambda = (2\alpha - 1)/(2\alpha)$, so that $\alpha(2 - \lambda) > 1$ and choose ρ so that (3.13) holds. Then

$$I_1(\rho, f') = \frac{1}{2\pi} \int_0^{2\pi} |f'(\rho e^{i\theta})| \, d\theta$$

$$\leqslant \left(\frac{1}{2\pi} \int_0^{2\pi} |f'(\rho e^{i\theta})|^2 |f(\rho e^{i\theta})|^{\lambda - 2} \, d\theta \right)^{\frac{1}{2}}$$

$$\times \left(\frac{1}{2\pi} \int_0^{2\pi} |f(\rho e^{i\theta})|^{2 - \lambda} \, d\theta \right)^{\frac{1}{2}}, \quad (3.14)$$

by Schwarz's inequality. We now take $r_0 = \frac{1}{2}$ in (3.10) and noting that $r \geqslant \frac{1}{2}$, $r \geqslant \rho$, we deduce

$$I_{2-\lambda}(\rho, f) \leqslant I_{2-\lambda}(r, f) \leqslant (1 - r_0)^{-\alpha(2-\lambda)} + \frac{p\Lambda}{r_0} \int_0^r (1 - t)^{-\alpha(2-\lambda)} \, dt$$

$$< A(p, \alpha)(1 - r)^{1 - \alpha(2-\lambda)},$$

since λ depends on α. We write $r_1 = 2r - 1$ so that $r_1 \leqslant \rho < r$ and deduce from (3.13), (3.14) and the above that

$$I_1(r_1, f') \leqslant I_1(\rho, f') \leqslant A(p, \alpha)(1 - r)^{-\frac{1}{2}[1 + \alpha\lambda - 1 + \alpha(2-\lambda)]}$$

$$= A(p, \alpha)(1 - r)^{-\alpha} = 2^\alpha A(p, \alpha)(1 - r_1)^{-\alpha}.$$

Here r_1 may be any number such that $0 < r_1 < 1$. Now (3.12) follows from (3.8). This proves Theorem 3.3 if $C = 1$. In the general case we consider $f(z)/C$ instead of $f(z)$.

3.3.1. The case $\alpha \leqslant \frac{1}{2}$.

If $\alpha \leqslant \frac{1}{2}$ in (3.11) we cannot prove as much as (3.12), and, as we shall see, (3.12) is false in general for $\alpha < \frac{1}{2}$, even if $f(z)$ is bounded. We observe that if (3.11) holds for some $\alpha \leqslant \frac{1}{2}$, then it holds a fortiori for every $\alpha > \frac{1}{2}$, and so (3.12) gives $|a_n| = O(n^{-\frac{1}{2}+\epsilon})$ for every $\epsilon > 0$. We can, however, improve on this slightly as follows.

THEOREM 3.4. *Suppose that* $f(z) = \sum_0^\infty a_n z^n$ *is mean p-valent in* $|z| < 1$ *and satisfies* (3.11). *Then we have, if* $\alpha = \frac{1}{2}$,

$$|a_n| < A(p) C n^{-\frac{1}{2}} \log(n + 1) \quad (n \geqslant 1), \quad (3.15)$$

and if $\alpha < \frac{1}{2}$

$$|a_n| < A(p, \alpha) C n^{-\frac{1}{2}} [\log(n + 1)]^{\frac{1}{2}} \quad (n \geqslant 1). \quad (3.16)$$

We suppose again that $C = 1$. It follows from (3.11), applied with $\frac{1}{2}(1+r)$ instead of r, that given λ $(0 < \lambda \leqslant 2)$ and r $(0 < r < 1)$, we can find ρ such that $r < \rho < \frac{1}{2}(1+r)$ and

$$\frac{1}{2\pi} \int_0^{2\pi} |f'(\rho e^{i\theta})|^2 |f(\rho e^{i\theta})|^{\lambda-2} d\theta \leqslant \frac{p}{\lambda} 2^{\alpha\lambda+3} (1-r)^{-1-\alpha\lambda}$$

$$\leqslant \frac{16p}{\lambda} (1-r)^{-1-\alpha\lambda}.$$

Choosing $$\frac{1}{\lambda} = \max \left\{ \alpha \log \frac{1}{1-r}, \frac{1}{2} \right\},$$

so that $(1-r)^{-\alpha\lambda} \leqslant e$, we deduce

$$\frac{1}{2\pi} \int_0^{2\pi} |f'(\rho e^{i\theta})|^2 |f(\rho e^{i\theta})|^{\lambda-2} d\theta \leqslant \frac{8pe}{1-r} \left[\log \frac{1}{1-r} + 1 \right].$$

We take $r_0 = \frac{1}{2}$ in (3.10) and have for $\frac{1}{2} < \rho < 1$

$$\frac{1}{2\pi} \int_0^{2\pi} |f(\rho e^{i\theta})|^{2-\lambda} d\theta \leqslant 2^{\alpha(2-\lambda)} + 2p(2-\lambda) \int_{\frac{1}{2}}^{\rho} (1-t)^{-\alpha(2-\lambda)} dt$$

$$\leqslant 2 + 2p(2-\lambda) \int_{\frac{1}{2}}^{\rho} (1-t)^{-2\alpha} dt.$$

The right-hand side is less than $A(p, \alpha)$ if $\alpha < \frac{1}{2}$, and so in this case (3.14) yields

$$I_1(r, f') \leqslant I_1(\rho, f') \leqslant A(p, \alpha) \left[\frac{1}{1-r} \log \frac{1}{1-r} \right]^{\frac{1}{2}} \quad (\tfrac{1}{2} \leqslant r < 1).$$

Putting $r = n/(n+1)$, we now deduce (3.16) from (3.8).

If $\alpha = \frac{1}{2}$, then

$$\int_{\frac{1}{2}}^{\rho} (1-t)^{-2\alpha} dt = \log \left[\frac{1}{2(1-\rho)} \right] \leqslant \log \frac{1}{1-r} \quad (\tfrac{1}{2} \leqslant r < 1),$$

and so we obtain from (3.14)

$$I_1(r, f') \leqslant \frac{A(p, \alpha)}{(1-r)^{\frac{1}{2}}} \left(\log \frac{1}{1-r} \right) \quad (\tfrac{1}{2} \leqslant r < 1),$$

and again (3.15) follows from (3.8).

3.4. A counter-example. The results of Theorem 3.4 are the strongest that are known, even for univalent functions, though they are probably not best possible. However, even if $f(z)$ is mean p-valent with p as small as we please and continuous in $|z| \leqslant 1$ nothing stronger than

$$|a_n| = o(n^{-\frac{1}{2}}) \tag{3.17}$$

is true in general. To see this let n_k be a rapidly increasing sequence of integers and put

$$f(z) = 1 + \sum_{n=1}^{\infty} a_n z^n,$$

where $a_n = \epsilon 2^{-k} n^{-\frac{1}{2}}$, if $n = n_k$ $(k = 1, 2, \ldots)$, $a_n = 0$ otherwise.

If λ_n is any pre-assigned sequence of positive numbers, tending to zero as $n \to \infty$, however slowly, we let $n_k \to \infty$ so rapidly with k, that

$$\lambda_{n_k} < \frac{\epsilon}{2^k} \quad (k = 1, 2, \ldots).$$

Thus for the infinite sequence of values of n given by $n = n_k$ we have

$$a_n > \frac{\lambda_n}{n^{\frac{1}{2}}}. \tag{3.18}$$

On the other hand, we have for $|z| \leqslant 1$,

$$|f(z) - 1| \leqslant \sum_{n=1}^{\infty} |a_n| \leqslant \epsilon \sum_{k=1}^{\infty} 2^{-k} = \epsilon,$$

so that the series for $f(z)$ converges uniformly and so $f(z)$ is continuous in $|z| \leqslant 1$. Also the area, with due count of multiplicity, of the image of $|z| < 1$ by $f(z)$ is

$$\int_0^1 r\, dr \int_0^{2\pi} |f'(r e^{i\theta})|^2 d\theta = \pi \sum_{n=1}^{\infty} n |a_n|^2 = \pi \sum_{k=1}^{\infty} \epsilon^2 2^{-2k} = \frac{\pi \epsilon^2}{3}$$

(see § 1.3). If $\pi W(R)$ denotes the amount of this area which lies over $|w| < R$, then $W(R) = 0$ if $R < 1 - \epsilon$, $W(R) < \pi \epsilon^2 / 3$ otherwise. If we choose $\epsilon < \frac{1}{2}$, it follows that

$$\frac{W(R)}{\pi R^2} < \frac{\pi \epsilon^2}{3\pi(\frac{1}{2})^2} = \frac{4\epsilon^2}{3} \quad (0 < R < \infty),$$

and the right-hand side can be made as small as we please by choosing ϵ small enough. Thus $f(z)$, which is continuous in $|z| \leqslant 1$, can be made mean p-valent there with p as small as we please. Clearly, for any function $f(z)$ bounded and mean p-valent in $|z| < 1$, $\sum_{1}^{\infty} n |a_n|^2$ converges and so (3.17) holds. Nevertheless, (3.18) shows that nothing stronger than this need be true.

It is pertinent to ask whether more than (3.17) is true, for instance, for bounded univalent functions. The answer to this question is not known, but Littlewood [2] has given a rather recondite example of such a function, which does not satisfy

$$|a_n| = O(n^{-1+A}) \quad (n \to \infty),$$

where A is a positive absolute constant. Thus the implication from (3.11) to (3.12) breaks down even for univalent functions if $\alpha \leqslant A$.

3.5. Coefficients of general mean p-valent functions.
Theorems 2.5, 3.3 and 3.4 give immediately

THEOREM 3.5. *Suppose that* $f(z) = \sum_{0}^{\infty} a_n z^n$ *is mean p-valent in* $|z| < 1$. *Then we have for* $1 \leqslant n < \infty$

$$|a_n| < A(p) \mu_p n^{2p-1} \quad (p > \tfrac{1}{4}), \tag{3.19}$$

$$|a_n| < A |a_0| n^{-\frac{1}{2}} \log (n+1) \quad (p = \tfrac{1}{4}), \tag{3.20}$$

$$|a_n| < A(p) |a_0| \left[\frac{\log (n+1)}{n} \right]^{\frac{1}{2}} \quad (0 < p < \tfrac{1}{4}), \tag{3.21}$$

where $A(p)$ *depends only on p and* $\mu_p = \max_{\nu \leqslant p} |a_\nu|$.

For by Theorem 2.5 we may write $C = A(p) \mu_p$, $\alpha = 2p$ in Theorems 3.3 and 3.4. The order of magnitude in (3.19), which is due to Biernacki [1] for p-valent functions, and to Spencer [2] for mean p-valent functions, is best possible. The example of the last section shows that the right-hand sides in (3.20), (3.21) cannot at any rate be replaced by $\epsilon_n n^{-\frac{1}{2}}$, where ϵ_n is a fixed sequence which tends to zero as $n \to \infty$.

3.6. Growth and omitted values. We give next an application of Theorem 3.4 to a problem raised by Dvoretzky [1]. We show that if a univalent function $f(z)$ omits a set of values which is fairly dense in the plane, then the effect on the coefficients is nearly as strong as if $f(z)$ is bounded.†

THEOREM 3.6. *Suppose that* $f(z) = \sum\limits_{0}^{\infty} a_n z^n$ *is univalent in* $|z| < 1$ *and maps* $|z| < 1$ $(1, 1)$ *conformally onto a domain* D *in the w plane. Let $d(R)$ be the radius of the largest disc,* $|w - w_0| < \rho$, *whose centre lies on* $|w_0| = R$ *and which lies entirely in D. Then if for some constant* $\alpha < \frac{1}{4}$ *we have*

$$d(R) \leqslant \alpha R \quad (0 < R < \infty), \tag{3.22}$$

then we have

$$|a_n| < A(\alpha) |a_0| \left[\frac{\log(n+1)}{n} \right]^{\frac{1}{2}} \quad (n = 1, 2, \ldots). \tag{3.23}$$

Let $z_0 = r e^{i\theta}$ and

$$\phi(z) = f\left(\frac{z_0 + z}{1 + \bar{z}_0 z} \right) = b_0 + b_1 z + \ldots.$$

Then $\phi(z)$ is univalent in $|z| < 1$ and maps $|z| < 1$ onto D. Hence there exists w_1 outside D, such that

$$|w_1 - b_0| \leqslant \alpha |b_0|.$$

Thus

$$\frac{\phi(z) - b_0}{b_1} = z + \ldots.$$

is univalent in $|z| < 1$ and omits the value $(w_1 - b_0)/b_1$. Now Theorem 1.2 gives

$$\left| \frac{w_1 - b_0}{b_1} \right| \geqslant \frac{1}{4},$$

and so $|b_1| \leqslant 4\alpha |b_0|$, i.e.

$$(1 - r^2) |f'(r e^{i\theta})| \leqslant 4\alpha |f(r e^{i\theta})|.$$

We deduce $\quad \dfrac{\partial}{\partial r} \log |f(r e^{i\theta})| \leqslant \dfrac{4\alpha}{1 - r^2} \quad (0 < r < 1),$

---

† For generalizations to mean p-valent and other functions see Hayman [3].

and integrating with respect to r, we obtain

$$\left|\frac{f(r\,e^{i\theta})}{f(0)}\right| \leqslant \left(\frac{1+r}{1-r}\right)^{2\alpha}.$$

Since $\alpha < \frac{1}{4}$ this gives

$$M(r,f) < \sqrt{2}\,|a_0|\,(1-r)^{-2\alpha} \quad (0 < r < 1),$$

and now Theorem 3.6 follows from Theorem 3.4, with 2α instead of α.

If $0 < \lambda \leqslant 1$, then the function

$$w = f(z) = \left(\frac{1+z}{1-z}\right)^{\lambda}$$

maps $|z| < 1$ $(1,1)$ conformally onto the angle $|\arg w| < \frac{1}{2}\pi\lambda$ and satisfies $d(R) = R\sin\left(\frac{1}{2}\pi\lambda\right)$. It does not satisfy (3.23) if $\lambda > \frac{1}{2}$. Thus in (3.22) $\frac{1}{4}$ cannot at any rate be replaced by any number greater than $\sin\left(\frac{1}{4}\pi\right) = \frac{1}{2}\sqrt{2}$.

3.7. Power series with k-fold symmetry.† Suppose that

$$f_k(z) = z + a_{k+1}z^{k+1} + a_{2k+1}z^{2k+1} + \dots$$

is univalent in $|z| < 1$; then so is

$$f(z) = [f_k(z^{1/k})]^k = z + b_2 z^2 + \dots,$$

and conversely. By applying the upper bound

$$M(r,f) \leqslant \frac{r}{(1-r)^2}$$

of Theorem 1.3, we obtain

$$M(r,f_k) \leqslant \frac{r}{(1-r^k)^{2/k}},$$

with equality only if $f_k(z) = z(1 - z^k e^{i\theta})^{-2/k}$. By applying Theorem 3.3 we can then deduce that if $k = 1$, 2, or 3

$$|a_{nk+1}| < A(k)\,n^{(2/k)-1} \quad (n \geqslant 1),$$

and this gives the correct order of magnitude. In fact Littlewood and Paley [1] first developed the ideas of Theorem 3.3 for this

---

† The rest of the chapter may be omitted on a first reading.

particular case. If $f_k(z)$ is p-valent, a similar argument shows that

$$| a_{nk+1} | < A(p,k)\, \mu_p\, n^{(2p/k)-1}, \qquad (3.24)$$

provided that $1 \leqslant k < 4p$ (Robertson [1]) and the proof was modified so as to apply to mean p-valent functions by Spencer [2, 3]. For this latter case (3.24) breaks down when $k > 4p$, as is shown by a suitable form of the examples in § 3.4.

For this type of proof it is essential that *all* the coefficients vanish, except those whose suffixes form an arithmetic progression of common difference k. We proceed to prove (3.24) under somewhat weaker assumptions, basing ourselves on the fact that the functions $f_k(z)$ satisfy

$$| f_k[r\, e^{i(\theta + 2\pi i \nu/k)}] | = | f_k(r\, e^{i\theta}) | \qquad (0 \leqslant \nu \leqslant k-1).$$

3.7.1. We need the following preliminary result:

THEOREM 3.7. *Suppose that* $f(z) = \sum_0^\infty a_n z^n$ *is mean p-valent in* $|z| < 1$ *and that there exist* $k \geqslant 2$ *points* $z_1', z_2', \ldots, z_k'$ *on* $|z| = r$, *where* $0 < r < 1$, *such that*

$$\text{(i)} \quad | z_i' - z_j' | \geqslant \delta \quad (1 \leqslant i < j \leqslant k),$$

and $\qquad$ (ii) $\quad | f(z_i') | \geqslant R \quad (1 \leqslant i \leqslant k).$

Then we have $\qquad R < A(p)\, \mu_p\, \delta^{2p(1/k-1)}(1-r)^{-2p/k}.$

We suppose that $\qquad \delta > 4^{p+2}(1-r). \qquad (3.25)$

For if this is false, we have by Theorem 2.5

$$R \leqslant M(r, f) < A(p)\, \mu_p (1-r)^{-2p}$$

$$\leqslant A(p)\, \mu_p (1-r)^{-2p/k} \left(\frac{4^{p+2}}{\delta} \right)^{2p-2p/k}$$

$$< A(p)\, \mu_p (1-r)^{-2p/k}\, \delta^{2(p/k)-2p},$$

so that Theorem 3.7 holds.

At least one of the annuli

$$1 - 4^{-\nu}\delta < | z | < 1 - 4^{-(\nu+1)}\delta \quad (1 \leqslant \nu \leqslant [p]+1)$$

is free from zeros of $f(z)$, since $f(z)$ has $q \leqslant p$ zeros in $|z| < 1$. For such a value of ν we choose

$$r_1 = 1 - \tfrac{1}{2}4^{-\nu}\delta = 1 - \delta_0,$$

say, and note that $f(z)$ has no zeros in $1 - 2\delta_0 < |z| < 1 - \tfrac{1}{2}\delta_0$. Using (3.25) we deduce further that

$$\frac{\delta}{8} \geqslant \delta_0 = \tfrac{1}{2}4^{-\nu}\delta > \tfrac{1}{2}4^{p+2-\nu}(1-r) \geqslant 2(1-r), \qquad (3.26)$$

since $1 \leqslant \nu \leqslant p+1$, and hence that $r_1 < r$.

Suppose now that

$$z_j' = r\,e^{i\theta_j} \quad (1 \leqslant j \leqslant k).$$

We write $$z_j = r_1 e^{i\theta_j} \quad (1 \leqslant j \leqslant k),$$

and apply Theorem 2.6 with $R_1 = M(r_1, f)$, R_2 the number R of Theorem 3.7 (ii), and δ_0 instead of δ. This gives

$$\delta_j = \frac{\delta_0 - |z_j' - z_j|}{\delta_0} = \frac{1 - r}{\delta_0} \quad (1 \leqslant j \leqslant k).$$

By construction the circles $|z - z_j| < \tfrac{1}{2}\delta_0$ contain no zeros of $f(z)$. Also the circle $|z - z_j| < \delta_0$ contains the point z_j' and has diameter $2\delta_0 \leqslant \tfrac{1}{4}\delta$ by (3.26). In view of hypothesis (i) of Theorem 3.7 it now follows that the circles $|z - z_j| < \delta_0$ $(1 \leqslant j \leqslant k)$ are non-overlapping and Theorem 2.6 is applicable. We obtain

$$k\left[\log \frac{A(p)\,\delta_0}{1-r}\right]^{-1} \leqslant 2p\left[\log\left(\frac{R_2}{eR_1}\right)\right]^{-1},$$

$$\frac{R_2}{eR_1} < \left\{\frac{A(p)\,\delta_0}{1-r}\right\}^{2p/k} < A(p)\left(\frac{\delta}{1-r}\right)^{2p/k}$$

by (3.26).† Since $r_1 = 1 - \delta_0$ Theorem 2.5 gives further

$$R_1 = M(r_1, f) < A(p)\,\mu_p\,\delta_0^{-2p} = A(p)\,\mu_p\left(\frac{2 \cdot 4^\nu}{\delta}\right)^{2p} < A(p)\,\mu_p\,\delta^{-2p},$$

and Theorem 3.7 follows on writing R instead of R_2.

---

† We may assume $R_2 < eR_1$. Since $\delta/(1-r) \geqslant 16$ by (3.26), our result is trivial otherwise.

3.8. Power series with gaps. It is easy to see that the theorems quoted in §3.7 follow at once from Theorems 2.5, 3.3 and 3.7. We can, however, prove more.

THEOREM 3.8. *Suppose that*

$$f(z) = \sum_0^{N-1} a_n z^n + a_N z^N + a_{N+k} z^{N+k} + a_{N+2k} z^{N+2k} + \cdots$$

is mean p-valent in $|z| < 1$. Then

$$M(r, f) < A(p, k, N) \mu_p (1-r)^{-2p/k} \quad (0 < r < 1), \quad (3.27)$$

and if $1 \leqslant k < 4p$ we have further

$$|a_n| < A(p, k, N) \mu_p n^{(2p/k)-1} \quad (n = 1, 2, \ldots). \quad (3.28)$$

Write $\qquad g(z) = \sum_0^{N-1} a_n z^n, \quad h(z) = \sum_{n=0}^{\infty} a_{N+kn} z^{N+kn},$

and for $r \geqslant \frac{1}{2}$ choose θ_0 so that $|f(r e^{i\theta_0})| = M(r, f)$. Then if

$$\theta_\nu = \theta_0 + \frac{2\pi\nu}{k} \quad (0 \leqslant \nu \leqslant k-1),$$

we have $\qquad h(r e^{i\theta_\nu}) = h(r e^{i\theta_0}) \exp\left(\frac{2\pi i\nu N}{k}\right).$

Again for $|z| = r < 1$, we have from Theorem 3.5

$$|g(z)| \leqslant \sum_{n=0}^{N-1} |a_n| \leqslant \sum_{n=0}^{N-1} A(p) \mu_p n^{2p-1} \leqslant A(p, N) \mu_p$$

if $p \geqslant \frac{1}{2}$ and

$$|g(z)| \leqslant \sum_{n=0}^{N-1} A |a_0| = AN |a_0|$$

if $p < \frac{1}{2}$. Thus

$$|f(r e^{i\theta_\nu})| \geqslant |h(r e^{i\theta_\nu})| - |g(r e^{i\theta_\nu})|$$
$$\geqslant |h(r e^{i\theta_0})| - A(p, N) \mu_p$$
$$\geqslant M(r, f) - 2A(p, N) \mu_p \quad (0 \leqslant \nu \leqslant k-1).$$

We may therefore apply Theorem 3.7 with

$$\delta = |r e^{2\pi i/k} - r| \geqslant \frac{1}{2} |e^{2\pi i/k} - 1| = A(k),$$

and $\qquad R = M(r, f) - A(p, N) \mu_p,$

and obtain

$$M(r,f) < A(p,N)\mu_p + A(p)\mu_p A(p,k)(1-r)^{-2p/k},$$

and this gives (3.27). The inequality, proved on the assumption $r \geqslant \frac{1}{2}$, clearly remains valid if $r < \frac{1}{2}$, since $M(r,f)$ increases with r. Also (3.28) follows from (3.27) and Theorem 3.3. This proves Theorem 3.8.

3.8.1. If $k = 2$, we can sharpen Theorem 3.7 considerably.

THEOREM 3.9. *Suppose that* $f(z) = \sum_{0}^{\infty} a_n z^n$ *is mean p-valent in* $|z| < 1$ *and that* $a_n = 0$ *whenever* $n = bm + c$, *where* b, c *are fixed positive integers and* m *goes from* 1 *to* ∞. *Then*

$$M(r,f) < A(p,b,c)\mu_p(1-r)^{-p} \quad (0 < r < 1), \qquad (3.29)$$

and so, if $p > \frac{1}{2}$,

$$|a_n| < A(p,b,c)\mu_p n^{p-1} \quad (n \geqslant 1). \qquad (3.30)$$

As a special case we note that, if $f(z) = z + \sum_{n=2}^{\infty} a_n z^n$ is univalent in $|z| < 1$ and satisfies the above hypotheses, then

$$|a_n| < A(b,c).$$

This result seems to require the full force of the methods of the previous chapter. It would be interesting to determine whether more general sequences of zero coefficients result in all the others being uniformly bounded.

To prove Theorem 3.9, consider

$$f_\nu(z) = \sum_{m=0}^{\infty} a_{bm+\nu} z^{bm+\nu} \quad (0 \leqslant \nu \leqslant b-1),$$

and write $\omega = \exp[2\pi i/b]$. Then we have

$$f_\nu(z) = \frac{1}{b} \sum_{\mu=0}^{b-1} \omega^{-\mu\nu} f(\omega^\mu z).$$

In fact the coefficient of z^n on the right-hand side is

$$\frac{a_n}{b} \sum_{\mu=0}^{b-1} \omega^{-\mu\nu}\omega^{\mu n} = \frac{a_n}{b} \sum_{\mu=0}^{b-1} [\omega^{n-\nu}]^\mu,$$

and this is a_n or 0, according as $n - \nu$ is or is not a multiple of b.

Hence if ν is so chosen that $c - \nu$ is a multiple of b, we have from the hypothesis of Theorem 3.9 for $|z| < 1$

$$\left| \sum_{\mu=0}^{b-1} \omega^{-\mu\nu} f(\omega^\mu z) \right| = b \left| f_\nu(z) \right| \leqslant b \sum_{n=0}^{b+c} |a_n| \leqslant A(p,b,c) \mu_p = K,$$

say.

Choose now, for $\frac{1}{2} \leqslant r < 1$, z_0 so that

$$|z_0| = r, \quad |f(z_0)| = M(r,f).$$

If $M(r,f) \leqslant 2K$, then (3.29) follows. Otherwise we obtain

$$\left| \sum_{\mu=1}^{b-1} \omega^{-\mu\nu} f(\omega^\mu z_0) \right| \geqslant |f(z_0)| - \left| \sum_{\mu=0}^{b-1} \omega^{-\mu\nu} f(\omega^\mu z_0) \right|$$

$$\geqslant |f(z_0)| - K \geqslant \tfrac{1}{2} |f(z_0)| = \tfrac{1}{2} M(r,f).$$

Hence we can find μ $(1 \leqslant \mu \leqslant b-1)$ such that

$$|f(\omega^\mu z_0)| \geqslant \frac{M(r,f)}{2(b-1)},$$

and $\quad |\omega^\mu z_0 - z_0| = r \, |1 - \omega^\mu| \geqslant \tfrac{1}{2} |1 - \omega| = \sin\left(\frac{\pi}{2b}\right) \geqslant \frac{1}{b}.$

We can thus apply Theorem 3.7, with $k=2$, $z_1' = z_0$, $z_2' = \omega^\mu z_0$, $\delta = b^{-1}$ and $R = \tfrac{1}{2} M(r,f)/(b-1)$. We obtain

$$M(r,f) < 2(b-1) A(p) \mu_p b^p (1-r)^{-p},$$

and this proves (3.29) if $r \geqslant \frac{1}{2}$. Since $M(r,f)$ increases with r, the result for $0 \leqslant r < \frac{1}{2}$ also follows, and (3.30) follows from (3.29) and Theorem 3.3. Thus Theorem 3.9 is proved.

It is worth noting that the above argument does not require the full strength of our hypotheses. It would be sufficient to assume that on $|z| = r$

$$|f_\nu(z)| = \left| \sum_{m=0}^{\infty} a_{bm+\nu} z^{bm+\nu} \right| = O(1-r)^{-p} \quad \text{as} \quad r \to 1,$$

in order to obtain $M(r,f) = O(1-r)^{-p}$. A similar remark applies to Theorem 3.8.

CHAPTER 4

SYMMETRIZATION

4.0. Introduction. In this chapter we develop the theory of symmetrization in the form due to Pólya and Szegö [1] as far as it is necessary for our function-theoretic applications.

Given a domain D, we can, by certain types of lateral displacement called symmetrization, transform D into a new domain D^* having some aspects of symmetry. The precise definition will be given in § 4.5. Pólya and Szegö showed that while area for instance remains invariant under symmetrization, various domain constants such as capacity, inner radius, principal frequency, torsional rigidity, etc., behave in a monotonic manner.

We shall here prove this result for the first two of these concepts in order to deduce Theorem 4.9, the principle of symmetrization. If $f(z) = a_0 + a_1 z + \ldots$ is regular in $|z| < 1$, and something is known about the domain D_f of values assumed by $f(z)$, this principle allows us to assert that in certain circumstances $|a_1|$ will be maximal when $f(z)$ is univalent and D_f symmetrical. Applications of this result will be given in the last three sections of the chapter. Some of these will in turn form the basis of further studies of p-valent functions in Chapter 5. Some of these results can also be proved in another manner by a consideration of the transfinite diameter (Hayman [2]).

Since the early part of the chapter is used only to prove Theorem 4.9, the reader who is prepared to take this result for granted on a first approach may start with § 4.5 and then go straight to § 4.9.

We shall need to refer to Ahlfors [2] (which we shall denote by C.A.) for a number of results which space does not permit us to consider in more detail here. A good set-theoretic background is provided by C.A., Chapter 2, § 2, and we use generally the notation there given. We shall, however, in accordance with common English usage, call an open connected set a *domain* and not a region. The closure of a domain D will be denoted by $\bar{D}$. A domain D has connectivity n if its complement in the extended plane has exactly n components; if $n = 1$, D is simply connected, if $n = 2$ doubly connected, etc. (C.A. pp. 112 and 118). If the boundary of D consists of a finite number n of analytic simple closed curves (C.A. pp. 65 and 191), no two of which have common points, we shall call D an analytic domain.

We shall assume a right-handed system OX, OY of rectangular Cartesian axes in the plane. Points in the plane will be denoted in terms of their coordinates x, y either by (x, y) or by $z = x + iy$, whichever is more convenient. Accordingly, functions u will be written as either $u(z)$ or $u(x, y)$.

4.1. Lipschitzian functions.

Let E be a plane set and let $P(z)$ be a function defined on E. We shall say that $P(z)$ is *Lipschitzian* or *Lip* on E if there is a constant C such that

$$\left| P(z_1) - P(z_2) \right| \leqslant C \left| z_1 - z_2 \right| \tag{4.1}$$

whenever z_1, z_2 lie in E. It is clear that a Lip function is continuous, and further that if P, Q are Lip and bounded on E, then PQ is Lip on E.

Suppose that E is compact and that $P(z)$ is Lip in some neighbourhood of every point z_0 of E. Then P is Lip on E. For if not, we could find sequences of distinct pairs of points z_n, z_n' $(n \geqslant 1)$ on E such that

$$\frac{\left| P(z_n) - P(z_n') \right|}{\left| z_n - z_n' \right|} \to \infty \quad (n \to \infty). \tag{4.2}$$

By taking subsequences if necessary we may assume that $z_n \to z_0$, $z_n' \to z_0'$, where z_0, z_0' lie in E. If z_0, z_0' are distinct, we at once obtain a contradiction from our local hypothesis and (4.2), since P is bounded near any point of E. If $z_0 = z_0'$, then z_n, z_n' finally lie in that neighbourhood of z_0 where P is Lip and this again contradicts (4.2).

If $P(z) = P(x, y)$ is defined in a disc γ (C.A. p. 53) and has bounded partial derivatives there, then $P(x, y)$ is Lip in γ. Suppose first that $P(x, y)$ is real. Then if (x_0, y_0), $(x_0 + h, y_0 + k)$ both lie in γ, then either $(x_0, y_0 + k)$ or $(x_0 + h, y_0)$ also lies in γ. Suppose, for example, the former. Then if M is a bound for the absolute values of the partial derivatives in γ we have from the mean-value theorem

$$| P(x_0 + h, y_0 + k) - P(x_0, y_0) |$$

$$\leqslant | P(x_0 + h, y_0 + k) - P(x_0, y_0 + k) | + | P(x_0, y_0 + k) - P(x_0, y_0) |$$

$$= \left| h \left(\frac{\partial P}{\partial x} \right)_{(x_0 + \theta h, \, y_0 + k)} \right| + \left| k \left(\frac{\partial P}{\partial y} \right)_{(x_0, \, y_0 + \theta' k)} \right|$$

$$\leqslant M(| h | + | k |) \leqslant 2M \sqrt{(h^2 + k^2)}.$$

Thus P is Lip in γ. For complex P we can prove the corresponding result by considering real and imaginary parts.

It follows that if P has continuous partial derivatives in a domain containing a compact set E, then P is Lip on E. For in this case P has continuous partial derivatives in some neighbourhood and so bounded partial derivatives in some smaller neighbourhood of every point of E.

Conversely, we note that if $P(x, y)$ is Lip on a segment $a \leqslant y \leqslant b$ of the line $x = \text{constant}$, then P is an absolutely continuous function of y on this segment and so $\partial P / \partial y$ exists almost everywhere on the segment and is uniformly bounded.[†] Thus

$$P(x, b) - P(x, a) = \int_a^b \frac{\partial P(x, y)}{\partial y} \, dy. \tag{4.3}$$

4.2. The formulae of Gauss and Green.

We proceed to prove these formulae in the form in which we shall require them in the sequel.

LEMMA 4.1 (Gauss's formula). *Suppose that D is a bounded analytic domain in the plane and that its boundary γ is described so as to leave D on the left. Then if $P(x, y)$, $Q(x, y)$ are Lip in $\overline{D}$, we have*

$$\int_\gamma (P \, dx + Q \, dy) = \iint_D \left(\frac{\partial Q}{\partial x} - \frac{\partial P}{\partial y} \right) dx \, dy.$$

Let $z = \alpha(t)$ $(a \leqslant t \leqslant b)$ give an arc of γ. Then $\alpha(t)$ is a regular function of t and $\alpha'(t) \neq 0$. The tangent is parallel to OY at those points where $\alpha'(t)$ is pure imaginary, and this can be true only

---

† Burkill[1], Chapter IV.

at a finite number of points, since otherwise $\alpha'(t)$ would be identically pure imaginary and so this arc of γ would reduce to a straight line.† This is, however, impossible, since γ consists of a finite number of analytic closed curves. Thus there are only a finite number of tangents to γ which are parallel to OY, and we assume that these are

$$x = x_m \quad (1 \leqslant m \leqslant M),$$

where $x_1 < x_2 < \dots < x_M$.

For $x_{m-1} < \xi < x_m$, the line $x = \xi$ meets γ in $2n$ points

$$y = y_1(\xi), \quad \dots, \quad y = y_{2n}(\xi),$$

where n depends only on m and the $y_\nu(\xi)$ $(1 \leqslant \nu \leqslant 2n)$ are differentiable functions of ξ for $x_{m-1} < \xi < x_m$. Also the part D_m of D lying in $x_{m-1} < x < x_m$ consists of n domains

$$D_{m,\nu}\colon y_{2\nu-1}(x) < y < y_{2\nu}(x), \quad x_{m-1} < x < x_m \quad (1 \leqslant \nu \leqslant n),$$

since at the intersections of $x = \xi$ with γ, the line $x = \xi$ alternately enters and leaves D.

Consider now

$$I = \int_\gamma P(x, y)\, dx$$

taken along γ so as to keep D on the left. Then $dx > 0$ on the curves $y = y_{2\nu-1}(x)$ and $dx < 0$ on the curves $y = y_{2\nu}(x)$. Thus if I_m is the integral taken over those points of γ which lie in $x_{m-1} < x < x_m$, then

$$I_m = \int_{x_{m-1}}^{x_m} \sum_{\nu=1}^{n} \{P[x, y_{2\nu-1}(x)] - P[x, y_{2\nu}(x)]\}\, dx$$

$$= \int_{x_{m-1}}^{x_m} dx \sum_{\nu=1}^{n} \int_{y_{2\nu-1}}^{y_{2\nu}} -\frac{\partial P(x, y)\, dy}{\partial y} = \sum_{\nu=1}^{n} \iint_{D_{m,\nu}} -\frac{\partial P}{\partial y}\, dx\, dy$$

by (4.3), since $P(x, y)$ is Lip. Adding over the separate ranges $x_{m-1} < x < x_m$ and all the domains $D_{m,\nu}$ we obtain

$$\int_\gamma P\, dx = \iint_D -\frac{\partial P}{\partial y}\, dx\, dy.$$

† The real part of $\alpha'(t)$ is a regular function of t for $a \leqslant t \leqslant b$ and so has only isolated zeros or vanishes identically.

Similarly we prove

$$\int_\gamma Q\,dy = \iint_D \frac{\partial Q}{\partial x}\,dx\,dy,$$

and Lemma 4.1 follows.

We deduce

LEMMA 4.2 (Green's formula). *Suppose that D is a bounded analytic domain with boundary γ, that u is Lip in $\overline{D}$, and that v possesses continuous second partial derivatives near every point of $\overline{D}$. Then*

$$\int_\gamma u\frac{\partial v}{\partial n}\,ds = -\iint_D \left[u\left(\frac{\partial^2 v}{\partial x^2} + \frac{\partial^2 v}{\partial y^2}\right) + \left(\frac{\partial u}{\partial x}\frac{\partial v}{\partial x} + \frac{\partial u}{\partial y}\frac{\partial v}{\partial y}\right) \right] dx\,dy,$$

where $\partial/\partial n$ denotes differentiation along the normal into D, and ds denotes arc length along γ.

In fact u, $\partial v/\partial x$, $\partial v/\partial y$ are Lip in $\overline{D}$ in this case and hence so are $u(\partial v/\partial x)$, $u(\partial v/\partial y)$. Let (x_0, y_0) be a point of γ and let θ be the angle the tangent to γ makes with OX, where γ is described so as to keep D on the left. Then the normal into D makes an angle $\theta + \frac{1}{2}\pi$ with OX and

$$\frac{\partial v}{\partial n} = \lim_{h\to 0} \frac{v[x_0 + h\cos(\theta + \frac{1}{2}\pi), y_0 + h\sin(\theta + \frac{1}{2}\pi)] - v[x_0, y_0]}{h}$$

$$= \cos\theta\,\frac{\partial v}{\partial y} - \sin\theta\,\frac{\partial v}{\partial x}.$$

If ds is an element of length of γ at (x_0, y_0), its projections on OX, OY are $dx = ds\cos\theta$ and $dy = ds\sin\theta$. Thus

$$\frac{\partial v}{\partial n}\,ds = \frac{\partial v}{\partial y}\,dx - \frac{\partial v}{\partial x}\,dy.$$

We now apply Lemma 4.1 with $P = u(\partial v/\partial y)$, $Q = -u(\partial v/\partial x)$ and Lemma 4.2 follows.

4.3. Harmonic functions and the problem of Dirichlet.

A function $u(x, y)$ is harmonic in a domain D, if u has continuous second partial derivatives in D, which satisfy Laplace's equation

$$\nabla^2(u) \equiv \frac{\partial^2 u}{\partial x^2} + \frac{\partial^2 u}{\partial y^2} = 0.$$

It follows from this that $\phi(z) = \partial u/\partial x - i(\partial u/\partial y)$ has continuous partial derivatives which satisfy the Cauchy-Riemann equations, and so $\phi(z)$ is regular in D. Also in any disc with centre z_0 that lies in D, u is the real part of the regular function

$$f(z) = \int_{z_0}^z \phi(\zeta)\,d\zeta + u(z_0).$$

Thus u possesses continuous partial derivatives of all orders in D. Also, by the maximum modulus theorem applied to $\exp\{\pm f(z)\}$, u can have no local maximum or minimum in D unless u is constant in D.

Suppose now that D is a domain in the open plane and that $u(\zeta)$ is a continuous function of ζ, defined on the boundary γ of D considered in the extended plane. Thus, if D is unbounded, γ includes the point at infinity. The problem of Dirichlet consists in finding a function $u(z)$ continuous in $\overline{D}$, harmonic in D and coinciding with $u(\zeta)$ on γ. If u_1, u_2 are two functions satisfying these conditions, then $u_1 - u_2$ is harmonic in D, continuous in $\overline{D}$ and zero on γ, and so by the maximum principle $u_1 - u_2$ vanishes identically in D. Thus a solution to the problem of Dirichlet is unique, if it exists.

A solution does not always exist. If D consists of the annulus $0 < |z| < 1$, then any function bounded and harmonic in D can be extended to be harmonic also at $z = 0$ and so in $|z| < 1$. Thus boundary values assigned on $|z| = 1$ determine $u(0)$.

The question of existence can often be settled by means of the following result, for whose proof we must refer the reader to C.A. p. 198.

THEOREM 4.1. *Suppose that given any boundary point z_0 of D there exists $\omega(z)$, harmonic in D, continuous in $\overline{D}$ and positive in $\overline{D}$, except at z_0, where $\omega(z_0) = 0$. Then the problem of Dirichlet possesses a solution for any assigned continuous boundary values on the boundary of D.*

From this we deduce the following criterion:

THEOREM 4.2. *The problem of Dirichlet always possesses a solution for the domain D if, given any boundary point of $\overline{D}$, there exists an arc c of a straight line or circle containing z_0 and lying outside D.*

We shall deduce Theorem 4.2 from Theorem 4.1. Suppose first that z_0 is finite. Let the arc c have end-points z_0 and z_1. The function

$$\zeta = e^{i\alpha}\frac{z-z_0}{z_1-z}$$

maps the exterior of this arc c onto the ζ plane cut along a ray through the origin, so that $z=z_0$, z_1 correspond to $\zeta=0,\infty$. By suitably choosing α, we may arrange that the ray is the negative real axis. Then

$$w = 1 + \frac{\zeta^{\frac{1}{2}}-1}{\zeta^{\frac{1}{2}}+1}$$

maps this cut plane onto the circle $|w-1| < 1$. Also $z=z_0$ corresponds to $w=0$, and D maps onto a domain whose closure lies in $|w-1| \leqslant 1$, and such that only $z=z_0$ corresponds to $w=0$. Thus $\omega = \Re w$ is the harmonic function whose existence is required in Theorem 4·1.† If z_0 is infinite and some ray from a finite point to infinity lies outside D, the argument is similar. This completes the deduction. A plane domain satisfying the criterion of Theorem 4.2 will be called *admissible* (for the problem of Dirichlet).

4.4. The Dirichlet integral and capacity.

Historically Dirichlet attempted to base a proof of the existence of a solution to his problem on the problem of finding the minimum for the Dirichlet integral

$$I_D(u) = \iint_D \left[\left(\frac{\partial u}{\partial x}\right)^2 + \left(\frac{\partial u}{\partial y}\right)^2\right] dx\,dy,$$

for all functions having the assigned boundary values and satisfying certain smoothness conditions. In favourable circumstances this minimum is attained by the required harmonic function. However, even for continuous boundary values on the unit circle, the harmonic function inside the circle with these boundary values may have an infinite Dirichlet integral, so that the minimum problem has no solution.

We shall need the Dirichlet minimum principle only in a special case, where its validity can be proved without difficulty.

---

† A point of $\overline{D}$ on the arc through z_0, z_1 corresponds to a pair of complex conjugate values of w. Thus ω is uniquely defined and continuous even on this arc.

THEOREM 4.3. *Let D be an admissible domain in the open plane whose complement consists of a compact set E_1 and a closed unbounded set E_0, not meeting E_1. Let $v(z)$, $\omega(z)$ both be continuous in the extended plane, $0, 1$ on E_0, E_1 respectively and Lip on every compact subset of D. Suppose, further, that $\omega(z)$ is harmonic in D. Then*

$$I_D[v(z)] \geqslant I_D[\omega(z)] = \int_{\gamma_a} \left| \frac{\partial \omega}{\partial n} \right| ds,$$

where γ_a is the set $\{z: \omega(z) = a\}$, and a is any number such that $0 < a < 1$ and $\partial\omega/\partial x - i(\partial\omega/\partial y) \neq 0$ on γ_a. In this case γ_a consists of a finite number of analytic Jordan curves.

The system consisting of the domain D and the sets E_0, E_1 will be called a *condenser*, and $I_D[\omega(z)]$ will be called the *capacity* of the condenser for evident physical reasons. In the special case when E_0, E_1 are continua, so that D is doubly connected, a simple function-theoretic interpretation is possible. We may then map D $(1, 1)$ conformally onto an annulus $\Delta\{z: 1 < |z| < R\}$. (For a construction of the mapping function in terms of $\omega(z)$ see C.A. p. 202.) The Dirichlet integral is clearly invariant under this transformation so that Δ has the same capacity as D. The harmonic function satisfying our boundary-value problem in Δ is

$$\Omega(z) = \frac{\log |R/z|}{\log R} \quad \text{in } \Delta,$$

and hence the capacity of Δ is

$$\int_{|z|=\rho} \frac{1}{\log R} \frac{\rho \, d\theta}{\rho} = \frac{2\pi}{\log R}.$$

The quantity $\log R$ is frequently called the *modulus* of the doubly connected domain D. It is thus a multiple of the reciprocal of the capacity of D considered as a condenser. For our purposes it is, however, essential that E_0, E_1 need not be connected.

4.4.1. Proof of Theorem 4.3.

Let $\omega(z)$ be the function of Theorem 4.3. Then $\phi(z) = \partial\omega/\partial x - i(\partial\omega/\partial y)$ is regular and not identically zero in D, and so $\phi(z)$ vanishes at most on a countable set of points in D, which we shall call the branch points of $\omega(z)$. Suppose that $0 < a < 1$ and that the set γ_a, where $\omega(z) = a$, does not pass through any branch point of $\omega(z)$. Clearly γ_a is compact and lies in D by our hypothesis. Also near any point z_0 of γ_a ω is the real part of a regular function

$$w = f(z) = \omega + i\omega_1 = a + ib + a_1(z - z_0) + \dots.$$

Here $a_1 \neq 0$, since $\phi(z) \neq 0$ on γ_a. Thus the inverse function $z = f^{-1}(w)$ is also regular and the part of γ_a near z_0 is the image by a regular function of a straight line segment, i.e. an analytic Jordan arc. If we continue along this arc we must finally return to our starting point, since otherwise γ_a would possess a limit point not of the kind described above. Thus γ_a consists of analytic Jordan curves, no two of which have common points. There can only be a finite number of these curves, since otherwise there would be a point of γ_a any neighbourhood of which meets an infinite number of curves. This again contradicts the local behaviour of γ_a established above.

Suppose again that γ_a, γ_b contain no branch points, that $0 < a < b < 1$, and that $D_{a,b} = \{z : a < \omega(z) < b\}$. Then $D_{a,b}$ consists of a finite number of bounded analytic domains, of which γ_a, γ_b form the complete boundary. Also if $\partial/\partial n$ denotes differentiation along the normal into $D_{a,b}$, then $\partial\omega/\partial n > 0$ on γ_a and $\partial\omega/\partial n < 0$ on γ_b. Now on applying Green's formula to the regions of $D_{a,b}$ separately and adding we obtain

$$\int_{\gamma_a} \frac{\partial\omega}{\partial n} ds + \int_{\gamma_b} \frac{\partial\omega}{\partial n} ds = 0,$$

and since $\partial\omega/\partial n$ has constant sign on γ_a, γ_b, we deduce

$$\int_{\gamma_a} \left| \frac{\partial\omega}{\partial n} \right| ds = \int_{\gamma_b} \left| \frac{\partial\omega}{\partial n} \right| ds = I,$$

say. Also a second use of Green's formula gives

$$(b - a) I = -\int_{\gamma_a} \omega \frac{\partial\omega}{\partial n} ds - \int_{\gamma_b} \omega \frac{\partial\omega}{\partial n} ds = I_{D_{a,b}}[\omega(z)].$$

Since every point of D belongs to some $D_{a,b}$, we deduce, on making $a \to 0$, $b \to 1$,

$$I_D[\omega(z)] = \lim_{a \to 0, b \to 1} I_{D_{a,b}}[\omega(z)] = I.$$

Suppose now that $v(z)$ is the function of Theorem 4.3 and set $h = v - \omega$. Then $h(z)$ is continuous in the plane and vanishes on E_0, E_1 and so at infinity. Hence, given $\epsilon > 0$, the set

$$E = \{z : |h(z)| \geqslant \epsilon\}$$

is a compact subset of D. Let a_0, b_0 be the lower and upper bounds of $\omega(z)$ on E, so that $0 < a_0 \leqslant b_0 < 1$. If $0 < a < a_0$, $b_0 < b < 1$, then γ_a, γ_b do not meet E and so $|h(z)| < \epsilon$ on γ_a, γ_b. Now h is Lip on $\bar{D}_{a,b}$ so that Green's formula gives

$$\left| \iint_{D_{a,b}} \left[\frac{\partial h}{\partial x} \frac{\partial \omega}{\partial x} + \frac{\partial h}{\partial y} \frac{\partial \omega}{\partial y} \right] dx\, dy \right| = \left| \int_{\gamma_a + \gamma_b} h \frac{\partial \omega}{\partial n} ds \right|$$

$$\leqslant \epsilon \int_{\gamma_a + \gamma_b} \left| \frac{\partial \omega}{\partial n} \right| ds = 2\epsilon I.$$

Thus

$$I_{D_{a,b}}(h, \omega) = \iint_{D_{a,b}} \left[\frac{\partial h}{\partial x} \frac{\partial \omega}{\partial x} + \frac{\partial h}{\partial y} \frac{\partial \omega}{\partial y} \right] dx\, dy \to 0 \quad (a \to 0,\ b \to 1).$$

Also

$$I_{D_{a,b}}(v) = I_{D_{a,b}}(\omega) + I_{D_{a,b}}(h) + 2 I_{D_{a,b}}(h, \omega).$$

Making $a \to 0$, $b \to 1$ we deduce $I_D(v) = I_D(\omega) + I_D(h)$. This proves the inequality of Theorem 4.3 and completes the proof of that theorem. We note also that equality is possible only if $I_D(h) = 0$. In this case it is not difficult to see that $h(z)$ vanishes identically so that $v(z)$ must coincide with $\omega(z)$. We shall not, however, make any use of this.

4.4.2. Capacity decreases with expanding domain. We make an immediate deduction from Theorem 4.3, which has independent interest:

THEOREM 4.4. *Let E_0, E_1; E_0', E_1' be pairs of closed sets characterizing two condensers as in Theorem 4.3. Let I, I' be their capacities and suppose that $E_0' \subset E_0$, $E_1' \subset E_1$. Then $I' \leqslant I$.*

Let $\omega(z)$ be the potential function corresponding to the first condenser as in Theorem 4.3. If $0 < a < b < 1$, we define a function $v(z)$ as follows: $v(z) = 0$ where $\omega(z) \leqslant a$, $v(z) = 1$ where $\omega(z) \geqslant b$, and

$$v(z) = \frac{\omega(z) - a}{b - a} \quad \text{in} \quad D_{a,b}.$$

Then $v(z)$ is Lip in the plane and $v = 0, 1$ on E_0', E_1' respectively. Thus if D' is the complement of E_0' and E_1', Theorem 4.3. gives

$$I' \leqslant I_{D'}(v) = - \int_{\gamma_a + \gamma_b} v \frac{\partial v}{\partial n} ds = \frac{1}{b-a} I_D[\omega(z)] = \frac{I}{b-a}.$$

Making $a \to 0$, $b \to 1$, we obtain $I' \leqslant I$ as required.

4.5. Symmetrization.

We shall now consider the process of symmetrization introduced in the middle of last century by Steiner and developed by Pólya and Szegö [1].

Let O be an arbitrary open set in the open plane. We shall define a symmetrized set O^* in two ways as follows.

4.5.1. Steiner symmetrization.

In this case we symmetrize with respect to a straight line, which we can take to be the x axis of Cartesian coordinates. For any real ξ, the line $x = \xi$

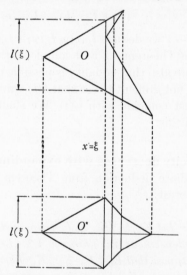

Fig. 4. Steiner symmetrization.

meets O in a set of mutually disjoint open intervals of total length $l(\xi)$ say, where $0 \leqslant l(\xi) \leqslant \infty$. The symmetrized set is to be the set

$$O^* = \{(x, y) : |y| < \tfrac{1}{2} l(x)\}.$$

4.5.2. Circular (Pólya) symmetrization.† In this case

we symmetrize with respect to a half-line or ray, which we take
to be the line $\theta = 0$ of polar coordinates r, θ. We define the symmetrized set O^* as follows. Consider the intersection of O with
the circle $r = \rho$. If this intersection includes the whole circle or
is null, then the intersection of O^* with $r = \rho$ is also to include the

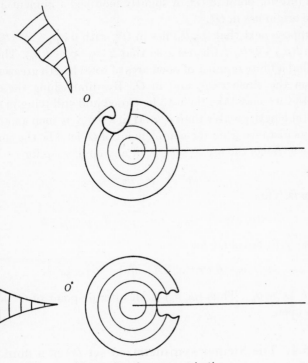

Fig. 5. Circular symmetrization.

whole circle, or to be null respectively. Otherwise let O meet
the circle $r = \rho$ in a set of open arcs of total length $\rho l(\rho)$, where
$0 < l(\rho) \leqslant 2\pi$. Then O^* is to meet the circle $r = \rho$ in the single arc

$$|\theta| < \tfrac{1}{2}l(\rho).$$

We proceed to discuss some properties of symmetrization.
Unless the contrary is stated results are true for either kind of
symmetrization.

† Pólya[1].

4.5.3. The symmetrized set O^* is open.

We prove this only for circular symmetrization. The proof for Steiner symmetrization is similar.

Suppose that (ρ_0, π) lies in O^*. Then O contains the whole circle $r = \rho_0$. Since O is open, O therefore contains some annulus $\rho_0 - \delta < r < \rho_0 + \delta$, and this annulus also lies in O^*. Thus (ρ_0, π) is an interior point of O^*. A slightly modified argument applies if the origin lies in O^*.

Suppose next that (ρ_0, θ_0) lies in O^*, with $0 < \rho_0 < \infty$, $|\theta_0| < \pi$. Then $l(\rho_0) > 2|\theta_0|$. Choose x so that $2|\theta_0| < x < l(\rho_0)$. Then we can find a finite number of open arcs of total length greater than $\rho_0 x$ on the circle $r = \rho_0$ and in O. By diminishing these arcs slightly, we may take them to be closed and still lying in O and of total length greater than $\rho_0 x$. If $\alpha_\nu \leqslant \theta \leqslant \beta_\nu$ is such an arc and δ_ν is its distance from the complement of O, let δ be the smallest of the δ_ν. Then the arc $\alpha_\nu \leqslant \theta \leqslant \beta_\nu$ of any circle $r = \rho$ for

$$\rho_0 - \delta < \rho < \rho_0 + \delta$$

lies in O. Thus

$$l(\rho) \geqslant x > 2|\theta_0| \quad \text{for} \quad \rho_0 - \delta < \rho < \rho_0 + \delta,$$

and so (r, θ) lies in O^* for

$$\rho_0 - \delta < r < \rho_0 + \delta \quad \text{and} \quad |\theta| < \tfrac{1}{2}x,$$

where $\tfrac{1}{2}x > |\theta_0|$. Thus (ρ_0, θ_0) is an interior point of O^* and so O^* is open.

4.5.4. The Steiner symmetrized set D^* of a domain D is either the whole plane, or a simply connected admissible domain.

Suppose that the lines $x = \xi_1, \xi_2$, where $\xi_1 < \xi_2$ meet D^* and so D. Then $x = \xi$ meets D for $\xi_1 < \xi < \xi_2$, since otherwise D could be expressed as the union of the disjoint non-null open subsets lying in the half-planes $x < \xi$ and $x > \xi$ respectively. Thus the segment $\xi_1 \leqslant x \leqslant \xi_2$ of the real axis lies in D^* in this case. Two points (ξ_1, η_1), (ξ_2, η_2) in D^* can be joined in D^* along the polygonal arc (ξ_1, η_1), $(\xi_1, 0)$, $(\xi_2, 0)$, (ξ_2, η_2). Thus D^* is a domain.

Suppose that next D^* is not the whole plane, so that $l(\xi) < +\infty$

for at least one ξ. Then the complement of D^* meets $x = \xi$ in the pair of rays $|y| > \frac{1}{2}l(\xi)$. Thus the condition for admissibility in Theorem 4.2 is satisfied for every point (ξ, η) in the complement of D^* and also at ∞. Further, every point in the complement of D^* can be joined to infinity by a ray, and so this complement is connected in the extended plane. Thus D^* is simply connected and is admissible.

The results for circular symmetrization are not so simple. If, for instance, D consists of an annulus $\rho_1 < r < \rho_2$ except for the single point (ρ, π), then D^* coincides with D and so is neither simply connected nor admissible.

4.5.5. The circularly symmetrized set of a domain D is a domain D^*. If D is simply connected, so is D^*. If D is admissible or D^* is simply connected, D^* is admissible or D^* is the whole plane. The proof that D^* is connected and is a domain is similar to that for Steiner symmetrization. If D so simply connected and contains the circle $r = \rho$, then D must contain the interior of this circle, since otherwise the complement of D would contain points inside and outside the circle (e.g. infinity) but no points on the circle. Hence if D contains $r = \rho$, D and D^* also contain the disc $r < \rho$ in this case. Let ρ_0 be the upper bound of all ρ for which this holds. If $\rho_0 = +\infty$, D and D^* contain the whole open plane. Otherwise the complement of D contains some point of every circle $r = \rho \geqslant \rho_0$, and so the complement of D^* contains the ray $r \geqslant \rho_0$, $\theta = \pi$.

Any boundary point of D^* on this ray and the point at infinity clearly satisfy the criterion of admissibility. If (ρ, θ_0) is any other point in the complement of D^*, then $\rho \geqslant \rho_0$, $|\theta_0| < \pi$, and the arc $|\theta_0| \leqslant \theta \leqslant \pi$ of the circle $r = \rho$ lies in the complement of D^* and joins (ρ, θ_0) to the ray $\theta = \pi$, $r \geqslant \rho_0$ in this complement. Thus the complement of D^* is connected and D^* is simply connected. The criterion of admissibility is also satisfied in all cases for every boundary point of D^* not on the line $\theta = \pi$. The above argument shows that if D^* is simply connected the criterion is also satisfied for boundary points (ρ, π).

It remains to show that if D is an admissible domain, so is D^*. By the above argument it suffices to establish the criterion of

Theorem 4.2 for the boundary points (ρ_0, π) of D^*, and for ∞. Since D is admissible the complement of D contains a ray, which must meet every sufficiently large circle $r = \rho$. Thus (ρ, π) lies in the complement of D^* for large ρ, and so this complement contains a ray $r > \rho$, $\theta = \pi$. Thus our criterion is satisfied at ∞. Suppose next that (ρ_0, π) is a boundary point of D^*. If an arc $\alpha \leqslant \theta \leqslant \beta$ of $r = \rho_0$ lies in the complement of D, then an arc of $r = \rho_0$ bisected by (ρ_0, π) lies in the complement of D^* and the criterion is satisfied. If not, let (ρ_0, θ) be a frontier point of D. Since D is admissible there is an arc of a straight line or circle containing (ρ_0, θ) but lying outside D and so by our hypothesis not on $r = \rho_0$. This arc must meet $r = \rho$ for all ρ either in some interval $\rho_0 - \delta \leqslant \rho \leqslant \rho_0$ or $\rho_0 \leqslant \rho \leqslant \rho_0 + \delta$. Thus the interval $[\rho_0 - \delta, \rho_0]$ or $[\rho_0, \rho_0 + \delta]$ of $\theta = \pi$ lies in the complement of D^* and contains (ρ_0, π), and so D^* is admissible.

4.6. Symmetrization of functions.

Let $u(z)$ be a function, real, continuous and bounded in the plane. We symmetrize u to obtain a new function $u^*(z)$ by simultaneously symmetrizing all the sets $D_a = \{z : u(z) > a\}$ $(-\infty < a < +\infty)$. These sets are open since u is continuous.

More precisely, let D_a^* be the symmetrized set of D_a with respect to some straight line or ray. For any point z in the plane we define $u^*(z)$ as the least upper bound of all a for which z lies in D_a^*.

In practice we shall be concerned only with the case where $u(z)$ is non-negative and vanishes continuously at infinity. In this case the sets D_a are bounded for $a > 0$ and their closures $\overline{D}_a$ are compact. Now a function $u(z)$ continuous on a compact set E is uniformly continuous on E (C.A. p. 64). In other words if $\Omega(\delta)$ is the upper bound of $|u(z_1) - u(z_2)|$ for z_1, z_2 on E and $|z_1 - z_2| \leqslant \delta$, then $\Omega(\delta)$ is finite and $\Omega(\delta) \to 0$ as $\delta \to 0$. The quantity $\Omega(\delta)$ is called *the modulus of continuity of* $u(z)$. Clearly $u(z)$ is Lip on E if and only if $\Omega(\delta) \leqslant C\delta$ for some positive C and $0 < \delta < \infty$.

4.6.1. Symmetrization decreases the modulus of continuity.

We shall be able to show that $u^*(z)$ is 'at least as continuous as $u(z)$' by proving

THEOREM 4.5. *Suppose that, with the above definitions of* $u(z)$, $u^*(z)$, *the set* $E = \{z: a \leqslant u(z) \leqslant b\}$ *is bounded and that* $u(z)$ *has modulus of continuity* $\Omega(\delta)$ *on* E. *Let* $E^* = \{z: a \leqslant u^*(z) \leqslant b\}$. *Then* $u^*(z)$ *is continuous on* E^* *with modulus of continuity* $\Omega^*(\delta) \leqslant \Omega(\delta)$. *In particular if* u *is Lip on* E, u^* *is Lip on* E^*.

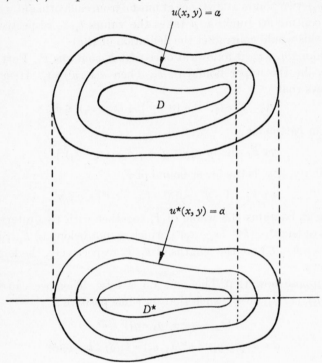

Fig. 6. Symmetrization of functions.

We shall prove this result for Steiner symmetrization. The proof for circular symmetrization is similar, but the details are a little more complicated. We need

LEMMA 4.3. *With the above hypotheses let* $l(x, t)$ *denote the total length of the* y *intervals on the line* $x = $ *constant where* $u(x, y) > t$. *Suppose that* $a \leqslant t_2 < t_1 - \Omega(\delta) \leqslant b - \Omega(\delta)$ *and* $|x_2 - x_1| \leqslant \delta$. *Then*

$$l(x_2, t_2) \geqslant l(x_1, t_1) + 2\sqrt{[\delta^2 - (x_2 - x_1)^2]}.$$

Let F_1 be the set of all y such that $u(x_1, y) > t_1$ and let F_2 be the set of y for which $u(x_2, y) > t_2$. Then

$$y_2 \in F_2 \quad \text{if} \quad y_1 \in F_1 \quad \text{and} \quad (x_2 - x_1)^2 + (y_2 - y_1)^2 \leqslant \delta^2.$$

For otherwise we could find, on the line joining (x_1, y_1) and (x_2, y_2), two points at distance at most δ from each other at which the continuous function u takes the values t_1, t_2 respectively, and this would contradict the definition of $\Omega(\delta)$.

Taking $y_2 = y_1$, it follows at once that F_2 contains F_1. Further, let y' be the upper bound of F_1. Then $u(x_1, y') = t_1$. Hence it follows that

$$u(x_2, y) > t_2 \quad \text{if} \quad (y - y')^2 + (x_2 - x_1)^2 \leqslant \delta^2,$$

and in particular

$$y \in F_2 \quad \text{if} \quad y' \leqslant y \leqslant y' + \sqrt{[\delta^2 - (x_2 - x_1)^2]}.$$

Similarly, if y'' is the lower bound of F_1

$$y \in F_2 \quad \text{if} \quad y'' - \sqrt{[\delta^2 - (x_2 - x_1)^2]} \leqslant y \leqslant y''.$$

Thus F_2 contains the whole of F_1 together with two intervals, each of length $\sqrt{[\delta^2 - (x_2 - x_1)^2]}$, which do not belong to F_1. Since $l(x_1, t_1)$, $l(x_2, t_2)$ are the lengths of F_1, F_2 respectively, the lemma follows.

Suppose now that Theorem 4.5 is false. Then we can find (x_1, y_1), (x_2, y_2) on E^*, such that for some positive δ

$$(x_2 - x_1)^2 + (y_2 - y_1)^2 \leqslant \delta^2,$$

and $a \leqslant u^*(x_2, y_2) < u^*(x_1, y_1) - \Omega(\delta) \leqslant b - \Omega(\delta).$

Choose t_1, t_2 so that

$$u^*(x_2, y_2) < t_2 < t_1 - \Omega(\delta) < u^*(x_1, y_1) - \Omega(\delta).$$

Then it follows from Lemma 4.3 that

$$l(x_2, t_2) \geqslant l(x_1, t_1) + 2\sqrt{[\delta^2 - (x_2 - x_1)^2]}.$$

Also by definition of $u^*(x, y)$ we have

$$l(x_1, t_1) > 2 \, |y_1|, \quad l(x_2, t_2) \leqslant 2 \, |y_2|,$$

since (x_1, y_1) lies in $D_{t_1}^*$, but (x_2, y_2) does not lie in $D_{t_2}^*$. Thus

$$|y_2 - y_1| \geqslant |y_2| - |y_1| > \sqrt{[\delta^2 - (x_2 - x_1)^2]}$$

and we have the contradiction which proves our theorem.

4.7. Symmetrization of condensers. Consider a condenser satisfying the conditions of Theorem 4.3. Let $\omega(z)$ be the potential function of Theorem 4.3. We symmetrize the function $\omega(z)$, and let

$$E_i^* = \{z \colon \omega^*(z) = i\} \quad (i = 0, 1).$$

By Theorem 4.5 $\omega^*(z)$ is continuous and so E_0^*, E_1^* are closed. It also follows that the complement of E_0^* is the symmetrized set of the complement of E_0, and that E_1^* is obtained by symmetrizing just as in the definitions 4.5.1. and 4.5.2 except that open set and open interval must be replaced by closed set and closed interval.

The complementary set to E_0^*, E_1^* consists of the open set $D^* = \{z \colon 0 < \omega^*(z) < 1\}$, and this is again a domain. Suppose, for example, that we are symmetrizing with respect to a half-line $\theta = 0$. Let r_0, r_1 be the lower and upper bounds of r in D. Then every circle $r = \rho$ $(r_0 < \rho < r_1)$ meets D^* in a single symmetrical arc $\gamma(\rho)$ of one of the forms $|\theta| < l(\rho)$, $|\theta| \leqslant \pi$, $l(\rho) < |\theta| \leqslant \pi$ or in a pair of arcs $\gamma_+(\rho)$, $\gamma_-(\rho)$ given by $l_1(\rho) < |\theta| < l_2(\rho)$. One of the former cases certainly holds unless $r = \rho$ meets both E_0 and E_1, and so if ρ is sufficiently near r_0 or r_1. It follows from the openness of D^*, that if ρ is sufficiently near to ρ_0, $\gamma_+(\rho)$ and $\gamma_+(\rho_0)$ (or $\gamma(\rho)$ and $\gamma(\rho_0)$) can be joined by a straight line segment in D^*, and the connectedness of D^* follows.

We can further show just as in 4.5.4 and 4.5.5 that D^* is admissible, since D is admissible. We have thus obtained a new condenser, which is called the symmetrized condenser of the original condenser. We now prove the following result of Pólya and Szegö [1]:

THEOREM 4.6. *Suppose that a condenser C and its symmetrized condenser C^* have capacities I, I^* respectively. Then $I^* \leqslant I$.*

By Theorem 4.5 $\omega^*(z)$ is continuous in the plane, and since $\omega(z)$ is Lip on the set $\bar{D}_{a,b} = \{z \colon a \leqslant \omega(z) \leqslant b\}$ if $0 < a < b < 1$, $\omega^*(z)$ is Lip on $\bar{D}_{a,b}^* = \{z \colon a \leqslant \omega^*(z) \leqslant b\}$. On any compact set E^* in D^*, $\omega^*(z)$ has lower and upper bounds a, b, where $0 < a < b < 1$, and so $\omega^*(z)$ is Lip on E^*. Also $\omega^*(z) = 0, 1$ on E_0^*, E_1^* respectively. Thus we have by Theorem 4.3

$$I^* \leqslant I_{D^*}[\omega^*(z)], \quad I = I_D[\omega(z)].$$

To complete the proof of Theorem 4.6 it thus remains to prove that

$$I_{D^*}[\omega^*(z)] \leqslant I_D[\omega(z)]. \tag{4.4}$$

This result represents the basic tool of the Pólya-Szegö theory.

4.7.1. Symmetrization decreases the Dirichlet integral.

We shall prove the inequality (4.4) for circular symmetrization. The proof for Steiner symmetrization is similar and even a little simpler. We express the Dirichlet integral in polar coordinates:

$$I_D(\omega) = \int_0^\infty \rho \, d\rho \int_{\mathfrak{E}_\rho} \left[\left(\frac{\partial \omega}{\partial \rho} \right)^2 + \frac{1}{\rho^2} \left(\frac{\partial \omega}{\partial \theta} \right)^2 \right] d\theta.$$

Here $\mathfrak{E}_\rho = \{\theta : 0 < \omega(\rho \, e^{i\theta}) < 1\}$.

Consider the expression

$$J(\rho) = \int_{\mathfrak{E}_\rho} \left[\left(\frac{\partial \omega}{\partial \rho} \right)^2 + \frac{1}{\rho^2} \left(\frac{\partial \omega}{\partial \theta} \right)^2 \right] d\theta.$$

If ω is constant on every circle with centre the origin, then D must be an annulus $r_1 < \rho < r_2$ and D^* coincides with D and $\omega(z)$ with $\omega^*(z)$. We ignore this case. Otherwise, since $\omega(z)$ is harmonic in D, $\omega(z)$ can be constant and different from $0, 1$ on the circle $r = \rho$ only for isolated values of ρ,† which we may omit from the range of integration. If $\omega(z)$ is not constant on $r = \rho$, $\omega(z)$ cannot be constant in any interval of $\mathfrak{E}_\rho$, since $\omega(\rho, \theta)$ is an analytic function of θ in $\mathfrak{E}_\rho$. For the end-points of such an interval would have to lie outside $\mathfrak{E}_\rho$, and so $\omega(z)$ would be 0 or 1 in the interval contrary to the definition of $\mathfrak{E}_\rho$. We thus assume that $\omega(z)$ is not constant in any interval of $\mathfrak{E}_\rho$, and so $\partial \omega / \partial \theta = 0$ only at isolated points in $\mathfrak{E}_\rho$. The values of $\omega(z)$ at these points will be called stationary values of $\omega(z)$. They can have at most $0, 1$ as limit points. We can thus arrange the stationary values in order of magnitude as a sequence t_m, where the lower bound of m is finite or $-\infty$ and the upper bound of m is finite or $+\infty$. If 0 is the lower bound of $\omega(z)$ on $\mathfrak{E}_\rho$ and 0 is not a limit point of the t_m, we include 0 as the smallest t_m; similarly for the value 1. Let t_m, t_{m+1} be successive stationary values. Then there will be n open intervals in $\mathfrak{E}_\rho$ in which $\omega(\rho, \theta)$ increases from t_m to t_{m+1}

† Otherwise $\partial \omega / \partial \theta$ would be identically zero.

and n intervals where ω decreases from t_{m+1} to t_m, where n is a positive integer, and these intervals occur alternately on the circle $r = \rho$, since ω is continuous on the circle.

Let us denote these intervals in the order in which they occur on the circle $r = \rho$ by $T_{m,\nu}$ ($1 \leqslant \nu \leqslant 2m$). We suppose that $\omega(\rho, \theta)$ increases in $T_{m,\nu}$ for odd ν and decreases in $T_{m,\nu}$ for even ν. Also the totality of intervals $T_{m,\nu}$ for varying m, ν make up $\mathfrak{E}_\rho$, except for isolated points. Thus

$$J(\rho) = \sum_m \sum_{\nu=1}^{2n(m)} \int_{T_{m,\nu}} \left[\left(\frac{\partial \omega}{\partial \rho} \right)^2 + \frac{1}{\rho^2} \left(\frac{\partial \omega}{\partial \theta} \right)^2 \right] d\theta.$$

We now change the variable of integration from θ to $t = \omega(\rho, \theta)$ in each interval $T_{m,\nu}$. Since $\partial \omega / \partial \theta \neq 0$ in $T_{m,\nu}$ we have

$$\frac{\partial \omega}{\partial \rho} = - \frac{\partial \theta}{\partial \rho} \bigg/ \frac{\partial \theta}{\partial t}, \quad \frac{\partial \omega}{\partial \theta} = 1 \bigg/ \frac{\partial \theta}{\partial t}, \quad d\theta = \frac{\partial \theta}{\partial t} \, dt.$$

Thus if $\theta_\nu(t, \rho)$ denotes the value of θ in $T_{m,\nu}$ such that $\omega(\rho, \theta) = t$, we have

$$J(\rho) = \sum_m \int_{t_m}^{t_{m+1}} \sum_{\nu=1}^{2n(m)} \left\{ \frac{(\partial \theta_\nu / \partial \rho)^2}{|\partial \theta_\nu / \partial t|} + \frac{1}{\rho^2} \frac{1}{|\partial \theta_\nu / \partial t|} \right\} dt. \qquad (4.5)$$

If $t_m < t < t_{m+1}$, the set of intervals on $r = \rho$ where $\omega(\rho, \theta) > t$ is given by $\theta_{2\nu-1}(t) < \theta < \theta_{2\nu}(t)$ ($1 \leqslant \nu \leqslant n$). Thus the length of this set is given by $\rho l(\rho, t)$, where

$$l(\rho, t) = \sum_{\nu=1}^{n} [\theta_{2\nu}(t) - \theta_{2\nu-1}(t)].$$

Clearly $l(\rho, t)$ decreases strictly with increasing t in $t_m < t < t_{m+1}$. Thus $\omega^*(\rho, \phi)$ satisfies

$$\omega^*(\rho, \phi) = t, \quad \text{where} \quad \phi = \pm \tfrac{1}{2} l(\rho, t) \quad \text{if} \quad t_m < t < t_{m+1}.$$

Thus if $J^*(\rho)$ corresponds to $\omega^*(z)$ as $J(\rho)$ corresponds to $\omega(z)$ we have†

$$J^*(\rho) = \sum_m 2 \int_{t_m}^{t_{m+1}} \left\{ \frac{(\partial \phi / \partial \rho)^2}{|\partial \phi / \partial t|} + \frac{1}{\rho^2} \frac{1}{|\partial \phi / \partial t|} \right\} dt, \qquad (4.6)$$

where $\qquad \phi = \tfrac{1}{2}[(\theta_2 - \theta_1) + \ldots + (\theta_{2n} - \theta_{2n-1})].$

---

† In estimating the integral $I_{D^*}[\omega^*(z)]$ we may ignore the sets where $\omega^* = t_m$. For such a set either has zero area or else $\partial \omega^* / \partial \rho$, $\partial \omega^* / \partial \theta$ vanish almost everywhere on it.

Thus

$$\frac{\partial \phi}{\partial t} = \frac{1}{2} \sum_{\nu=1}^{2n} (-1)^{\nu} \frac{\partial \theta_{\nu}}{\partial t} = -\frac{1}{2} \sum_{1}^{2n} \left| \frac{\partial \theta_{\nu}}{\partial t} \right|, \quad \frac{\partial \phi}{\partial \rho} \leqslant \frac{1}{2} \sum_{1}^{2m} \left| \frac{\partial \theta_{\nu}}{\partial \rho} \right|.$$

Thus the arithmetic-harmonic mean theorem gives

$$\frac{2}{\left| \dfrac{\partial \phi}{\partial t} \right|} = \frac{4}{\displaystyle\sum_{1}^{2n} \left| \dfrac{\partial \theta_{\nu}}{\partial t} \right|} \leqslant \frac{4}{(2n)^2} \sum_{1}^{2n} \frac{1}{\left| \dfrac{\partial \theta_{\nu}}{\partial t} \right|}.$$

Also Schwarz's inequality gives $(\Sigma a)^2 \leqslant (\Sigma a^2/|b|)(\Sigma|b|)$ and so

$$\frac{2(\partial \phi/\partial \rho)^2}{|\partial \phi/\partial t|} \leqslant \frac{(\Sigma|\partial \theta_{\nu}/\partial \rho|)^2}{\Sigma|\partial \theta_{\nu}/\partial t|} \leqslant \Sigma \left\{ \frac{|\partial \theta_{\nu}/\partial \rho|^2}{|\partial \theta_{\nu}/\partial t|} \right\}.$$

Substituting these inequalities in the expressions (4.5) and (4.6) for $J(\rho)$ and $J^*(\rho)$ we deduce

$$J^*(\rho) \leqslant J(\rho),$$

and hence on integration with respect to ρ we deduce (4.4). This completes the proof of Theorem 4.6.

4.8. Green's function and the inner radius.

Suppose that D is a domain in the complex z plane, z_0 a point of D, and that there exists a function $g[z, z_0, D]$, continuous in the closed plane and harmonic in D except at z_0, vanishing outside D and such that

$$g[z, z_0, D] + \log|z - z_0|$$

remains harmonic at $z = z_0$. Then $g[z, z_0, D]$ is called the (classical) *Green's function* of D.

Clearly $g[z, z_0, D]$ is unique if it exists, for if $g_1(z)$ is another function with the same properties, then $g-g_1$ is harmonic in the whole of D, continuous in $\bar{D}$, and zero outside D, and so $g-g_1$ vanishes identically by the maximum principle.

If D is admissible and bounded $g[z, z_0; D]$ exists. For let $h(z, z_0)$ be harmonic in D and have boundary values $\log|z - z_0|$ on the boundary of D. Then clearly

$$g(z, z_0) = h(z, z_0) - \log|z - z_0|$$

is the required Green's function in D.

Suppose now that $g(z, z_0)$ exists. Then $g > 0$ in D, since otherwise g would have a minimum in D at a point other than z_0, and this is impossible. Since

$$g(z, z_0) + \log | z - z_0 |$$

remains harmonic at z_0, the limit

$$\gamma = \lim_{z \to z_0} g(z, z_0) + \log | z - z_0 |$$

exists. We write

$$\gamma = \log r_0$$

and call r_0 the *inner radius* of D at z_0.

To explain this terminology, suppose that D is simply connected and that

$$w = \psi(z) = z - a_0 + b_1(z - a_0)^2 + \dots$$

maps D (1, 1) conformally onto a circle $| w | < r_0$, so that $\psi(a_0) = 0$, $\psi'(a_0) = 1$. Then

$$g(z, a_0) = \log \frac{r_0}{| \psi(z) |}$$

is the Green's function of D at a_0, and the inner radius of D at a_0 is r_0. For clearly $g(z, a_0)$ is harmonic in D except at a_0 and as z approaches the boundary of D in any manner $| \psi(z) | \to r_0$ and so $g(z) \to 0$. Also near $z = a_0$ we have

$$g(z, a_0) = \log \frac{r_0}{| z - a_0 | | 1 + o(1) |} = \log \frac{1}{| z - a_0 |} + \log r_0 + o(1),$$

as required.

We also note that if

$$z = f(w) = a_0 + a_1 w + \dots$$

maps $| w | < 1$ onto D (1, 1) conformally, then

$$w = f^{-1}(z) = \frac{z - a_0}{a_1} + \dots$$

near $z = a_0$, so that $a_1 f^{-1}(z)$ has the properties of $\psi(z)$ in the above analysis. In particular, $| a_1 |$ is the inner radius of D at a_0 in this case. We remark specifically, however, that our definition of the inner radius applies also to multiply connected domains D, which cannot be mapped onto a circle, and even in the simply connected case we do not need to assume the existence of the mapping.

We note that if the domains D, D_1 possess Green's functions $g(z, z_0)$, $g_1(z, z_0)$, and if $D \subset D_1$, then the difference

$$g_1(z, z_0) - g(z, z_0)$$

is harmonic in D and non-negative on the boundary of D. Thus the difference is non-negative in D, and so if r, r_1 are the inner radii of D, D_1 at z_0, then $\log r_1 \geqslant \log r$ and so $r \leqslant r_1$. *Thus the inner radius increases with expanding domain.*

We accordingly define the inner radius r_0 at a point a_0 of an arbitrary domain D, containing a_0, as the least upper bound of the inner radius at a_0 of all domains containing a_0, contained in D, and possessing (classical) Green's functions. We shall have $0 < r_0 \leqslant \infty$.

4.8.1. The inner radius and conformal mapping. Our application of the inner radius to function theory derives from

THEOREM 4.7. *Suppose that $w = f(z) = a_0 + a_1 z + \dots$ is regular in $|z| < 1$ and takes there values w, which lie in a domain D having inner radius r_0 at a_0. Then $|a_1| \leqslant r_0$. Equality holds if $f(z)$ maps $|z| < 1$ $(1, 1)$ conformally onto D.*†

We discussed in the previous section the case when $f(z)$ maps $|z| < 1$ $(1, 1)$ conformally onto D. In the general case take $0 < \rho < 1$ and choose ρ so that $f'(z) \neq 0$ on $|z| = \rho$. This can be achieved by increasing ρ slightly if necessary. Let $C(\rho)$ be the curve which is the image of $|z| = \rho$ by $w = f(z)$. Thus $C(\rho)$ is a closed analytic curve which may cross itself. The set of values w assumed by $w = f(z)$ inside $|z| < \rho$ forms a domain $D(\rho)$ whose boundary consists of certain arcs of $C(\rho)$. If two such arcs touch at a point P, we add the interior of a small circle of centre P to $D(\rho)$. In this way we construct a domain D_0 which contains $D(\rho)$, is contained in D, and whose boundary consists of a finite number of analytic arcs no two of which touch each other.

---

† This is, in fact, the only case of equality. See, for example, Hayman[2] for the limiting case $a_0 = \infty$.

Thus D_0 satisfies the condition of admissibility of Theorem 4.2. Further, D_0 is bounded and so possesses a Green's function

$$g_\rho(w, a_0) = \log \left| \frac{1}{w - a_0} \right| + \log r(\rho) + o(1)$$

near $w = a_0$, where $r(\rho)$ is the inner radius of D_0 at a_0. Since D_0 lies in D, we have $r(\rho) \leqslant r_0$.

Consider now the function

$$h(z) = g_\rho(f(z), a_0) + \log \frac{|z|}{\rho}.$$

Then $h(z)$ is non-negative on $|z| = \rho$ and harmonic in $|z| \leqslant \rho$ except possibly at points where $f(z) = a_0$. At all such points except possibly the origin $h(z)$ becomes positively infinite. At the origin we have

$$h(z) = \log \left| \frac{1}{f(z) - a_0} \right| + \log r(\rho) + \log \frac{|z|}{\rho} + o(1)$$

$$= \log \frac{r(\rho)}{|a_1| \rho} + o(1).$$

Thus $h(z)$ remains harmonic at the origin and

$$h(0) = \log [r(\rho)/|a_1| \rho].$$

Thus $h(z) \geqslant 0$ in $|z| < \rho$, since otherwise $h(z)$ would have a negative minimum somewhere in $|z| < \rho$, and at such a point $h(z)$ would be harmonic, which is impossible. Thus

$$h(0) = \log \frac{r(\rho)}{|a_1| \rho} \geqslant 0,$$

$$|a_1| \leqslant \frac{r(\rho)}{\rho} \leqslant \frac{r_0}{\rho}.$$

Making $\rho \to 1$, we have the inequality of Theorem 4.7.

4.8.2. The inner radius and symmetrization. Theorem 4.7 becomes a powerful tool in the theory of functions, when combined with the following result of Pólya and Szegö [1]:

THEOREM 4.8. *Suppose that a_0 is a point of a domain D in the w plane and that D^* is obtained from D by symmetrizing with*

6

respect to a line or half-line passing through a_0. Let r_0, r_0^* be the inner radii of D, D^* at a_0. Then $r_0 \leqslant r_0^*$.†

Following Pólya and Szegö we shall deduce Theorem 4.8 from Theorem 4.6. Suppose first that D is bounded and admissible so that D has a Green's function $g(w) = g(w, a_0)$, satisfying

$$g(w) = \log \left| \frac{1}{w - a_0} \right| + \log r_0 + o(1)$$

near $w = a_0$. We have

$$\frac{\partial}{\partial r} g(a_0 + r e^{i\theta}) = \frac{-1}{r} + O(1) \quad (r \to 0).$$

Thus if K is sufficiently large, there is exactly one value of r for each θ such that $g(a_0 + r e^{i\theta}) = K$, and the set of these points $a_0 + r e^{i\theta}$ forms an analytic Jordan curve γ_K surrounding a_0. Let $r = |w - a_0|$. On γ_K we have

$$r = e^{-K}[1 + o(1)] r_0$$

for large K. In other words, given $\epsilon > 0$, we have for sufficiently large K

$$r_0 e^{-(K+\epsilon)} < r < r_0 e^{-(K-\epsilon)} \quad \text{on} \quad \gamma_K. \tag{4.7}$$

Consider now the condenser formed by keeping γ_K and its interior at unit potential and the boundary of D at zero potential. The corresponding potential function is $g(w)/K$, and hence by Theorem 4.3 the capacity of the condenser is

$$-\frac{1}{K} \int_{\gamma_{K'}} \frac{\partial g}{\partial n} ds,$$

where $0 < K' < K$. Since g is harmonic it follows from Lemma 4.2 that if r is small

$$-\int_{\gamma_K} \frac{\partial g}{\partial n} ds = -\int_{|w-a_0|=r} \frac{\partial g}{\partial r} ds = 2\pi r \left[\frac{1}{r} + O(1) \right] \to 2\pi \quad (r \to 0).$$

Thus the capacity of our condenser is $2\pi/K$.

Consider next the capacity $c(r)$ of the condenser formed on replacing γ_K by the circle $|w - a_0| = r$. By Theorem 4.4 we have $c(r) \geqslant 2\pi K^{-1}$ or $c(r) \leqslant 2\pi K^{-1}$ according as $|w - a_0| \leqslant r$ includes

† Jenkins[2] has shown that when D is simply connected $r_0 < r_0^*$ unless D^* coincides with D.

or is included in the interior of γ_K. Thus we have from (4.7) for large K

$$c(r_0 e^{-K-\epsilon}) < \frac{2\pi}{K} < c(r_0 e^{-K+\epsilon}),$$

or

$$\log\frac{r_0}{r} - \epsilon < \frac{2\pi}{c(r)} < \log\frac{r_0}{r} + \epsilon$$

for all sufficiently small r. Thus

$$\frac{1}{c(r)} = \frac{1}{2\pi}\log\frac{r_0}{r} + o(1) \quad (r \to 0).$$

If we now symmetrize our condenser with respect to a line or half-line through a_0, then the disk $|w-a_0| \leqslant r$, being already symmetrical, is unaltered. The domain D is replaced by its symmetrized domain D^*. If $c^*(r)$ denotes the capacity of the new condenser, we have by Theorem 4.6

$$\frac{1}{c(r)} \leqslant \frac{1}{c^*(r)}.$$

On applying the above asymptotic expression for $1/c(r)$ we deduce

$$\log\frac{r_0}{r} + o(1) \leqslant \log\frac{r_0^*}{r} + o(1),$$

and this gives $r_0 \leqslant r_0^*$.

We have thus proved Theorem 4.8 when D is bounded and admissible. For a general domain D in the open plane we argue as follows. Let r_0 be the inner radius of D at a_0. Let D_1 be a domain in D, having a Green's function $g(w, a_0)$ and inner radius r_1 at a_0, where $r_1 > r_0 - \epsilon$. Suppose that ϵ is so chosen that

$$\frac{\partial g}{\partial u} - i\frac{\partial g}{\partial v} \neq 0$$

on the curves $\gamma_\epsilon = \{w : g(w) = \epsilon\}$, where $w = u + iv$. Let D_2 be that domain in D_1, bounded by some of the analytic Jordan curves γ_ϵ, which contains a_0. Then D_2 is analytic and so admissible, the Green's function of D_2 is $g - \epsilon$, and the inner radius, r_2, of D_2 at a_0 is $r_1 e^{-\epsilon}$.

Thus if D^*, D_2^* are the symmetrized domains of D, D_2 and r_0^*, r_2^* their inner radii at a_0, we have $r_0^* \geqslant r_2^* \geqslant r_2 > e^{-\epsilon} (r_0 - \epsilon)$. Since ϵ is arbitrary, the inequality $r_0 \leqslant r_0^*$ of Theorem 4.8 is proved.

4.9. The principle of symmetrization.† We may combine Theorems 4.7 and 4.8 in the following result, which is in a form suitable for applications:

THEOREM 4.9. *Suppose that* $w = f(z) = a_0 + a_1 z + \dots$ *is regular in* $|z| < 1$, *and that* $D = D_f$ *is the domain of all values* w *assumed by* $w = f(z)$ *at least once in* $|z| < 1$. *Suppose further that the symmetrized domain* D^* *of* D *with respect to a line or ray through* a_0 *lies in a simply connected domain* D_0 *and that*

$$w = \phi(z) = a_0 + a_1' z + a_2' z^2 + \dots$$

maps $|z| < 1$ (1.1) *conformally onto* D_0. *Then* $|a_1| \leqslant |a_1'|$.

Let r, r^* and r_0 be the inner radii of D, D^* and D_0 at a_0. Then we have

$$|a_1| \leqslant r \leqslant r^* \leqslant r_0 = |a_1'|.$$

The first inequality follows from Theorem 4.7, as does the equation $|a_1'| = r_0$. Again we have $r \leqslant r^*$ by Theorem 4.8 and $r^* \leqslant r_0$, since, as was shown in § 4.8, the inner radius increases with expanding domain. This proves Theorem 4.9.

Using the work of Jenkins [2] and Hayman [2] quoted above, one can show that strict inequality holds in Theorem 4.9, unless D coincides with D_0 and D^*, and $f(z)$ with $\phi(z e^{i\lambda})$. However, we shall not need to use this result.

In Theorem 4.9 we may allow $f(z) = a_0 + a_1 z + \dots$ to vary over a class $\mathfrak{F}$ of functions for which a_0 is fixed and D_f^* remains inside D_0. Then if $\phi(z) \epsilon \mathfrak{F}$, Theorem 4.9 shows that $|a_1'|$ gives the exact upper bound for $|a_1|$, when $f(z)$ varies over the class $\mathfrak{F}$. We shall give some examples of this type of result, which may have independent interest, in the next two sections. Frequently the extremal functions $\phi(z)$ give also the exact upper bounds for $M(r, f)$ and $M(r, f')$ when $f(z) \epsilon \mathfrak{F}$, and in the last section of the chapter we shall give a proof of this in a typical case, which is important for Chapter 5.

† For the remaining results of this chapter see Hayman [1].

4.10. Applications of Steiner symmetrization.

We suppose that the hypotheses of Theorem 4.9 are satisfied and that a_0 is real, and we symmetrize with respect to the real axis $v = 0$ in the $w = u + iv$ plane. Let $\theta(u)$ be the total measure of the intervals in which D_f meets the line $u =$ constant. Then D^* is the domain
$$\{w: v < \tfrac{1}{2}\theta(u), \; -\infty < u < +\infty\}.$$

As we saw in § 4.5.4., D^* is necessarily simply connected in this case, and so we might choose $D_0 = D^*$ in Theorem 4.9, unless D^* is the whole plane. For by the Riemann mapping theorem (C.A. p. 172) any other simply connected domain may be mapped $(1, 1)$ conformally onto $|z| < 1$ by a function $\phi(z) = a_0 + a_1'z + \dots$, where a_0 is an assigned point of the domain.

We shall not appeal to this general result, but simply give two examples of the method where the extremal function $\phi(z)$ can easily be calculated explicitly so that numerical inequalities result. Evidently many other examples could be given.

THEOREM 4.10.[†] *Suppose that $f(z) = a_0 + a_1 z + \dots$ is regular in $|z| < 1$ and that D_f intersects each line $u =$ constant in the $w = u + iv$ plane in a set of intervals of total length at most l. Then*

$$|a_1| \leqslant \frac{2l}{\pi}.$$

Equality holds for $f(z) = a_0 + \dfrac{l}{\pi}\log\dfrac{1+z}{1-z}$, which maps $|z| < 1$ $(1, 1)$ conformally onto the strip $|v - \Im a_0| < \tfrac{1}{2}l$.

We may, without loss in generality, suppose a_0 real by subtracting an imaginary constant if necessary. We then symmetrize with respect to the real axis and note that D_f^* lies in the strip $|v| < \tfrac{1}{2}l$. Since

$$\phi(z) = a_0 + \frac{l}{\pi}\log\frac{1+z}{1-z} = a_0 + \frac{2l}{\pi}z + \dots$$

maps $|z| < 1$ $(1, 1)$ conformally onto this strip, the inequality $|a_1| \leqslant 2l/\pi$ follows from Theorem 4.9. This inequality is sharp, since $\phi(z)$ satisfies the hypotheses for $f(z)$.

† For univalent $f(z)$ this is an unpublished result of Rogosinski. For the general case see Hayman[1, 2].

As another example we prove

THEOREM 4.11. *Suppose that* $f(z) = a_0 + a_1 z + \ldots$ *is regular in* $|z| < 1$ *and that* D_f *meets the imaginary axis in the w plane in a set of intervals of total length at most l. Then*

$$|a_1| \leqslant [4|a_0|^2 + l^2]^{\frac{1}{2}}.$$

Equality holds if a_0 is real and $f(z)$ maps $|z| < 1$ $(1,1)$ *conformally onto the domain D_0 consisting of the w plane cut from* $-\frac{1}{2}il$ *to* $-i\infty$ *and from $\frac{1}{2}il$ to* $+i\infty$ *along the imaginary axis.*

Suppose first that a_0 is real. We may then symmetrize with respect to the real axis and then D_f^* lies in D_0. Thus if

$$\phi(z) = a_0 + a_1' z + \ldots$$

maps $|z| < 1$ onto D_0 we shall have $|a_1| \leqslant |a_1'|$. It remains to find a_1'. We note that

$$\zeta = z + \frac{1}{z}$$

maps $|z| < 1$ onto the closed ζ plane cut along the segment $(-2, 2)$ of the real axis. Thus

$$w = \frac{il}{\zeta}$$

maps $|z| < 1$ onto D_0. We now choose r so that $-1 < r < 1$ and put

$$iz = \frac{z_1 + r}{1 + rz_1}.$$

Thus we obtain a more general map of $|z_1| < 1$ onto D_0, given by

$$w = \frac{l(z_1 + r)(1 + rz_1)}{(1 + rz_1)^2 - (z_1 + r)^2} = \frac{lr}{(1 - r^2)} + \frac{l(1 + r^2)}{(1 - r^2)} z_1 + \ldots .$$

We may choose r so that $lr/(1 - r^2) = a_0$. Then

$$a_1' = \frac{l(1 + r^2)}{(1 - r^2)} = \sqrt{(4a_0^2 + l^2)},$$

and Theorem 4.11 follows for real a_0.

If, finally, $a_0 = \alpha + i\beta$, we consider $f(z) - i\beta$ instead of $f(z)$ and deduce

$$|a_1| \leqslant \sqrt{(4\alpha^2 + l^2)} \leqslant \sqrt{(4|a_0|^2 + l^2)}.$$

This completes the proof.

4.11. Applications of circular symmetrization. Circular

symmetrization is more powerful than Steiner symmetrization, and any result obtainable by the latter method can also be obtained by the former on taking exponentials, though this may be less direct. When we use circular symmetrization the domain D_f^* is not necessarily simply connected. On the other hand, D_f^* cannot reduce to the whole plane unless D_f does, whereas in Steiner symmetrization D_f^* consists of the whole plane as soon as D_f meets every line perpendicular to the line of symmetrization in a set of intervals of infinite total length.

We proceed to give some examples.

THEOREM 4.12. *Suppose that* $0 < \alpha < 2$, *that*

$$f(z) = a_0 + a_1 z + \dots$$

is regular in $|z| < 1$ *and that* D_f *meets each circle* $|w| = \rho$ $(0 < \rho < \infty)$ *in a set of arcs of total length* $\pi \alpha \rho$ *at most. Then* $|a_1| \leqslant 2\alpha |a_0|$, *with equality when* $f(z) = a_0 \left(\dfrac{1+z}{1-z} \right)^\alpha$.

We may suppose that a_0 is real and positive, since this may be achieved by considering $e^{-i\lambda} f(z)$ instead of $f(z)$. With this assumption we symmetrize D_f with respect to the positive real axis. Then D_f^* lies in the domain

$$D_0 = \{w \colon |\arg w| < \tfrac{1}{2}\alpha\pi,\ 0 < |w| < \infty\}.$$

The function

$$w = \phi(z) = a_0 \left(\frac{1+z}{1-z} \right)^\alpha = a_0(1 + 2\alpha z + \dots)$$

maps $|z| < 1$ onto D_0 and now Theorem 4.12 follows from Theorem 4.9.

Our key result for further applications is

THEOREM 4.13. *Let* $f(z) = a_0 + a_1 z + \dots$ *be regular in* $|z| < 1$, *and let* $R = R_f$ *be the least upper bound, supposed finite, of all* ρ *for which* D_f *contains the whole circle* $|w| = \rho$. *If there are no such* ρ, *we put* $R = 0$. *Then* $|a_1| \leqslant 4(|a_0| + R)$.

Equality holds when $a_0 \geqslant 0$ *and*

$$f(z) = a_0 + \frac{4(a_0 + R)z}{(1-z)^2},$$

which maps $|z| < 1$ *onto* D_0 *consisting of the* w *plane cut from* $-\infty$ *to* $-R$ *along the negative real axis.*

We again suppose $a_0 \geqslant 0$ and symmetrize with respect to the positive real axis. Then since D_f does not contain the whole circle $|w| = \rho$ for $\rho > R$, D_f^* does not contain the point $w = -\rho$ for $\rho > R$ and so D_f^* lies in D_0. Since

$$\phi(z) = a_0 + \frac{4(a_0 + R)z}{(1-z)^2} = a_0 + 4(a_0 + R)z + \ldots$$

evidently maps $|z| < 1$ onto D_0 (see § 1.1), Theorem 4.13 follows.

Various special cases of Theorem 4.13 are of interest. Thus if

$$f(z) = z + a_2 z^2 + \ldots \quad (|z| < 1),$$

we may take $a_0 = 0$, $a_1 = 1$ and obtain $R \geqslant \frac{1}{4}$. This is a sharp form of a theorem of Landau [1]. The result also includes Theorem 1.2 of Koebe-Bieberbach. In fact if $f(z)$ is, in addition, univalent, D_f is simply connected and so its complement is connected. Thus if this complement contains a point w_0, it must meet every circle $|w| = \rho$ for $\rho > |w_0|$, and so $|w_0| \geqslant R \geqslant \frac{1}{4}$. Thus D_f contains the disc $|w| < \frac{1}{4}$.

Again if $R = 0$, we obtain $|a_1| \leqslant 4|a_0|$, a limiting case of Theorem 4.12.

It is easy to obtain inequalities when in addition to the other hypotheses $f(z)$ is bounded above. As an example we state

THEOREM 4.14. *Suppose that* $f(z)$ *satisfies the hypotheses of Theorem 4.13 and that in addition* $|f(z)| < M$ *in* $|z| < 1$. *Then*

$$|a_1| \leqslant \frac{4(M - |a_0|)(M^2 + |a_0|R)(|a_0| + R)}{(M+R)^2 (M + |a_0|)}.$$

Equality holds when $a_0 > 0$, *and* $f(z)$ *maps* $|z| < 1$ (1, 1) *conformally onto* D_0, *consisting of* $|w| < M$, *cut from* $-M$ *to* $-R$ *along the negative real axis.*

The proof is left to the reader. Clearly D_f^* lies in D_0, and the determination of the map of $|z| < 1$ onto D_0, which gives the extremal value for $|a_1|$, is an elementary exercise in conformal mapping.

Finally, we obtain the following corollary of Theorem 4.13:

THEOREM 4.15. *Suppose that $w = f(z) = a_0 + a_1 z + \dots$ is regular in $|z| < 1$, that $0 < l < \infty$, and that the complement of D_f in the w plane contains a sequence of points*

$$u + i(v + nl) \quad (-\infty < n < +\infty),$$

where v may depend on u, for every real u. Then $|a_1| \leqslant 2l/\pi$. Equality holds for $f(z) = a_0 + \dfrac{l}{\pi} \log \left(\dfrac{1+z}{1-z} \right)$ as in Theorem 4.10.

Consider
$$g(z) = e^{2\pi f(z)/l} = e^{2\pi a_0/l} \left(1 + \frac{2\pi a_1}{l} z + \dots \right).$$

Then $g(z)$ is regular in $|z| < 1$, and since

$$\log g(z) \neq \frac{2\pi}{l} (u + iv + inl) \quad (-\infty < n < +\infty),$$

$$g(z) \neq \exp \left[\frac{2\pi}{l} (u + iv) \right].$$

Since u is arbitrary subject to $-\infty < u < +\infty$, we can for every $\rho > 0$ find a number w, such that $|w| = \rho$ and $g(z) \neq w$ in $|z| < 1$. Thus we can apply Theorem 4.13 to $g(z)$ with $R = R_g = 0$, and obtain

$$\left| \frac{2\pi a_1}{l} \right| \leqslant 4.$$

Equality holds for real a_0, when

$$f(z) = a_0 + \frac{l}{\pi} \log \left(\frac{1+z}{1-z} \right),$$

which maps $|z| < 1$ onto the strip $|v| < \frac{1}{2}l$. This proves Theorem 4.15.

A comparison of Theorems 4.10 and 4.15 illustrates the relative power of Steiner and circular symmetrization. It may be seen that Theorem 4.15, while containing Theorem 4.10 as a special case, is significantly stronger. For, under the hypotheses of Theorem 4.15, D_f may well meet each line $u = \text{constant}$ in a set of intervals of infinite length, provided a single sequence of points in arithmetic progression of common difference l is on the line and outside D_f. The results also suggest that the hypotheses of Theorem 4.15 might be further weakened to the assumption that D_f meets no line $u = \text{constant}$ in any *interval* of length greater than l. It is not known whether the inequality $|a_1| \leqslant 2l/\pi$ still holds under this assumption.

4.12. Bounds for $|f(z)|$ and $|f'(z)|$. The functions which give the maximum for $|a_1|$ under the hypotheses of Theorems 4.10–4.15 also give the maximum values for $M(r, f)$ and $M(r, f')$ for $0 < r < 1$, and we can even prove in some cases that they are essentially the only functions that do so. The proofs are similar and we confine ourselves to the case of Theorem 4.13, which has most applications.

THEOREM 4.16. *Suppose that $f(z)$ satisfies the hypotheses of Theorem 4.13. Then we have, for $|z| = r$ $(0 < r < 1)$,*

$$|f(z)| < |a_0| + \frac{4(|a_0| + R) r}{(1-r)^2}, \tag{4.8}$$

$$|f'(z)| \leqslant \frac{4(R + |f(z)|)}{1 - r^2} < \frac{4(|a_0| + R)(1+r)}{(1-r)^3}, \tag{4.9}$$

unless
$$f(z) = a_0 + \frac{4(R e^{i\lambda} + a_0) z e^{-i\theta}}{(1 - z e^{-i\theta})^2},$$

where θ, λ are real and $\arg a_0 = \lambda$ or $a_0 = 0$.

We suppose $|z| < 1$ and consider

$$\phi(z) = f\left(\frac{z_0 + z}{1 + \bar{z}_0 z}\right) = f(z_0) + (1 - |z_0|^2) f'(z_0) z + ...,$$

instead of $f(z)$. Then $D_\phi = D_f$, since ϕ is obtained from f by a map of the unit disc onto itself, and so we can apply Theorem 4.13 to $\phi(z)$ with the same value of R. This gives

$$(1 - |z_0|^2) |f'(z_0)| \leqslant 4(R + |f(z_0)|),$$

and, dropping the suffix, we obtain the left inequality of (4.9). We next write $z_0 = r e^{i\theta}$, $f(z_0) = T(r) e^{i\lambda(r)}$, where θ is fixed. Then

$$e^{i\theta} f'(r e^{i\theta}) = e^{i\lambda(r)} [T'(r) + i\lambda'(r) T(r)].$$

Thus we have
$$T'(r) \leqslant \frac{4(R + T(r))}{1 - r^2}, \tag{4.10}$$

and equality is possible only if $\lambda'(r) = 0$. We may write (4.10) as

$$d \log [R + T(r)] \leqslant d \left[2 \log \left(\frac{1+r}{1-r} \right) \right].$$

Thus
$$\psi(r) = \left(\frac{1-r}{1+r} \right)^{2'} [R + T(r)] \qquad (4.11)$$

cannot increase with increasing r. If $\psi(r_1) = \psi(r_2)$, where $0 \leqslant r_1 < r_2 < 1$, then $\psi(r)$ is constant for $r_1 \leqslant r \leqslant r_2$, and so equality holds in (4.10). Then $\lambda'(r) = 0$ in this range, so that

$$\lambda(r) = \lambda = \text{constant}.$$

This gives
$$e^{i\theta} f'(z) = \frac{4[R e^{i\lambda} + f(z)]}{1 - (z e^{-i\theta})^2}$$

for $z = r e^{i\theta}$ $(r_1 < r < r_2)$, and hence by analytic continuation throughout $|z| < 1$. Thus

$$\frac{d}{dz} \log [R e^{i\lambda} + f(z)] = \frac{4 e^{-i\theta}}{1 - (z e^{-i\theta})^2},$$

$$\frac{R e^{i\lambda} + f(z)}{R e^{i\lambda} + a_0} = \left(\frac{1 + z e^{-i\theta}}{1 - z e^{-i\theta}} \right)^2,$$

and since $\arg f(r e^{i\theta}) \equiv \lambda$, we must have $\arg a_0 = \lambda$ unless $a_0 = 0$. Thus in this case $f(z)$ reduces to one of the extremals given in Theorem 4.16. They clearly satisfy the hypotheses of that theorem and give equality in the inequalities (4.8) and (4.9), when $z = r e^{i\theta}$.

In all other cases the function $\psi(r)$ in (4.11) decreases strictly with increasing r for $0 < r < 1$, and in particular

$$\psi(r) < \psi(0) = R + |a_0|,$$

i.e.
$$|f(r e^{i\theta})| < (R + |a_0|) \left(\frac{1+r}{1-r} \right)^2 - R.$$

This proves (4.8), and by substitution we obtain the second inequality of (4.9). This completes the proof of Theorem 4.16.

Two special cases of Theorem 4.16 are worth noting. If $a_0 = 0$, then (4.8) gives

$$M(r, f) < \frac{4Rr}{(1-r)^2}.$$

We thus obtain the following sharp form of a theorem of Bohr [1]:

If $f(z)$ is regular in $|z| < 1$, $f(0) = 0$, and $M(r, f) = 1$ for some r such that $0 < r < 1$, then D_f includes a circle $|w| = \rho$, where

$$\rho > \frac{(1-r)^2}{4r},$$

except when
$$f(z) = \frac{(1-r)^2}{r} \frac{z\, e^{i\lambda}}{(1 - z\, e^{-i\theta})^2}.$$

For, except in this case,

$$R > \frac{(1-r)^2}{4r} M(r, f) = \frac{(1-r)^2}{4r},$$

and we may choose ρ so that $\tfrac{1}{4} r^{-1}(1-r)^2 < \rho < R$, and $|w| = \rho$ lies in D_f.

We also collect together the results when $R = 0$:

THEOREM 4.17. *Suppose that $f(z) = a_0 + a_1 z + \ldots$ is regular in $|z| < 1$ and that for each ρ $(0 \leqslant \rho < \infty)$ we can find $w = \rho\, e^{i\phi(\rho)}$ such that $f(z) \neq w$ in $|z| < 1$. Then unless $f(z) = a_0[(1 + z\, e^{i\theta})/(1 - z\, e^{i\theta})]^2$ for some real θ we have for $|z| = r$ $(0 < r < 1)$*

$$|a_0| \left(\frac{1-r}{1+r}\right)^2 < |f(z)| < |a_0| \left(\frac{1+r}{1-r}\right)^2, \tag{4.12}$$

$$|f'(z)| \leqslant \frac{4}{1-r^2} |f(z)| < 4 |a_0| \frac{1+r}{(1-r)^3}, \tag{4.13}$$

and $\left(\dfrac{1-r}{1+r}\right)^2 M(r, f)$ decreases strictly with increasing r $(0 < r < 1)$.

In this case we take $R = 0$ in Theorem 4.16, and (4.8) and (4.9) give the right-hand inequality of (4.12) and (4.13). Also $[f(z)]^{-1}$ satisfies the same hypotheses as $f(z)$, and we obtain the left-hand inequality in (4.12) by applying (4.8) to $[f(z)]^{-1}$ instead of $f(z)$.

To prove the last statement of Theorem 4.17, let $0 \leqslant r_1 < r_2 < 1$ and choose θ so that

$$|f(r_2\, e^{i\theta})| = M(r_2, f).$$

Then unless $f(z) \equiv a_0[(1+z e^{i\theta})/(1-z e^{i\theta})]^2$, the function $\psi(r)$ of (4.11) decreases strictly with increasing r and so

$$\left(\frac{1-r_2}{1+r_2}\right)^2 M(r_2, f) = \left(\frac{1-r_2}{1+r_2}\right)^2 |f(r_2 e^{i\theta})|$$

$$< \left(\frac{1-r_1}{1+r_1}\right)^2 |f(r_1 e^{i\theta})| \leqslant \left(\frac{1-r_1}{1+r_1}\right)^2 M(r_1, f).$$

This completes the proof of Theorem 4.17.

CHAPTER 5

CIRCUMFERENTIALLY MEAN
p-VALENT FUNCTIONS

5.0. Introduction. We consider again functions $f(z)$ regular and not constant in a domain Δ, define $n(w) = n(w, \Delta, f)$ as the number of roots of the equation $f(z) = w$ in Δ and write as in (2.3)

$$p(R) = p(R, \Delta, f) = \frac{1}{2\pi} \int_0^{2\pi} n(R\,e^{i\phi})\,d\phi.$$

In Chapters 2 and 3 we considered functions satisfying the condition

$$\int_0^R p(\rho)\,d(\rho^2) \leqslant pR^2 \quad (R > 0),$$

where p is a positive number. In what follows we shall call such functions *areally mean p-valent*. We shall consider in this chapter some consequences of the more restrictive hypothesis

$$p(R) \leqslant p \quad (R > 0). \tag{5.1}$$

Functions satisfying this condition were introduced by Biernacki [2] and are generally called *circumferentially mean p-valent*. For the purpose of this chapter we shall call mean p-valent in Δ any function regular in Δ and satisfying (5.1).

We first prove some sharp inequalities restricting the growth of functions mean p-valent in $|z| < 1$ and such that either $f(z) \neq 0$, or $f(z)$ has a zero of order p at $z = 0$, where p is a positive integer, basing ourselves on Theorems 4.13 and 4.17 of the last chapter.

In the second part of the chapter we prove some regularity theorems for general mean p-valent functions in $|z| < 1$. We shall show in fact that if $f(z) = \sum_0^\infty a_n z^n$ is such a function, then

$$\alpha = \lim_{r \to 1-} (1 - r)^{2p} M(r, f)$$

exists, and if further $p > \frac{1}{4}$, then also

$$\lim_{n \to \infty} \frac{|a_n|}{n^{2p-1}} = \frac{\alpha}{\Gamma(2p)}.$$

A generalization to positive powers of mean p-valent functions and functions of the type $\sum\limits_{n=0}^{\infty} a_{kn+\nu} z^{kn+\nu}$ will also be proved. As an application of these results, we shall prove that if

$$f(z) = z + a_2 z^2 + \dots$$

is mean 1-valent in $|z| < 1$, then $|a_n| \leqslant n$ for $n > n_0(f)$.

5.1. Functions without zeros. In this section we prove†

THEOREM 5.1. *Suppose that $f(z) = a_0 + a_1 z + \dots$ is mean p-valent and $f(z) \neq 0$ in $|z| < 1$. Then*

$$|a_1| \leqslant 4p |a_0|.$$

Further, unless $f(z) = a_0[(1 + z e^{i\theta})/(1 - z e^{i\theta})]^{2p}$ for a real θ, we have for $|z| = \rho$ $(0 < \rho < 1)$

$$|a_0| \left(\frac{1-\rho}{1+\rho} \right)^{2p} < |f(z)| < |a_0| \left(\frac{1+\rho}{1-\rho} \right)^{2p}$$

and $\qquad |f'(z)| \leqslant \dfrac{4p}{1-\rho^2} |f(z)| < \dfrac{4 |a_0| p(1+\rho)^{2p-1}}{(1-\rho)^{2p+1}}.$

Also $[(1-\rho)/(1+\rho)]^{2p} M(\rho, f)$ decreases strictly with increasing ρ $(0 < \rho < 1)$, and so approaches a limit α_0 as $\rho \to 1$ where $\alpha_0 < |a_0|$.

We need two lemmas:

LEMMA 5.1. *If $f(z)$ is mean p-valent in a domain Δ and $\eta > 0$, then $\psi(z) = [f(z)]^\eta$ is mean (ηp)-valent in Δ, provided that $\psi(z)$ is single-valued there.*

Put $\psi(z) = W = R e^{i\Phi}$, $f(z) = w = r e^{i\phi}$, so that $W = w^\eta$. Thus we may write $R = r^\eta$, $\Phi = \eta \phi$. If on passing along a level curve $|f(z)| = r = $ constant, $\phi = \arg f(z)$ increases strictly by an amount $\delta \phi$, then this level curve contributes $\dfrac{1}{2\pi} \delta \phi$ to $p(r, \Delta, f)$. Also the corresponding contribution to $p(R, \Delta, \psi)$ where $R = r^\eta$, is $\dfrac{\eta}{2\pi} \delta \phi$. We call $\delta \phi$ the variation of $\arg f(z)$ on the arc.

† Hayman[1, 2, 4].

We may express the set of level curves $|f(z)| = r$ in Δ as the union of a finite or enumerable set of Jordan arcs, no two of which have more than end-points in common and on each of which $\arg f(z)$ varies monotonically. Then $2\pi p(r, \Delta, f)$ is the sum of the variations of $\arg f(z)$ on these Jordan arcs, and so we obtain by addition

$$p[R, \Delta, \psi] = \eta p[r, \Delta, f] \leqslant \eta p,$$

where $R = r^\eta$. This proves the lemma.

LEMMA 5.2. *If $f(z)$ is mean 1-valent in $|z| < 1$, then there exists a number $l = l_f \geqslant 0$, with the following property. If $|w| < l$ then the equation $f(z) = w$ has exactly one root in $|z| < 1$, while if $R \geqslant l$, we can find $w_R = R e^{i\phi}$, such that $f(z) \neq w_R$ in $|z| < 1$.*

Let $n(w)$ be the number of roots of the equation $f(z) = w$ in $|z| < 1$. We note that the sets $\{w : n(w) \geqslant K\}$ in the w plane are open for every finite positive K. In fact if $n(w_0) \geqslant K$, then there exist a finite number of distinct points $z_1, z_2, \ldots, z_q$ in $|z| < 1$ at which $f(z_\nu) = w_0$ with multiplicity K_ν and $\sum_{\nu=1}^{q} K_\nu \geqslant K$. We can then, given a small positive δ, choose ϵ so small that for $0 < |w - w_0| < \epsilon$ the equation $f(z) = w$ has exactly K_ν roots in $|z - z_\nu| < \delta$.[†] If δ is sufficiently small all these roots are distinct and so

$$n(w) \geqslant \sum_{\nu=1}^{q} K_\nu \geqslant K \quad \text{if} \quad |w - w_0| < \epsilon.$$

Thus the set where $n(w) \geqslant K$ is open.

It follows that if $n(R e^{i\phi_0}) > 1$, then $n(R e^{i\phi}) \geqslant 2$ in some range $|\phi - \phi_0| < \epsilon$. Since also

$$\frac{1}{2\pi} \int_0^{2\pi} n(R e^{i\phi}) \, d\phi \leqslant 1,$$

by hypothesis, we deduce that $n(R e^{i\phi}) < 1$, i.e. $n(R e^{i\phi}) = 0$, for some other value of ϕ. Thus if $n(R e^{i\phi}) \geqslant 1$ $(0 \leqslant \phi \leqslant 2\pi)$, then $n(R e^{i\phi}) = 1$ in this range. In this case the circle $|w| = R$ corresponds $(1, 1)$ continuously to a set γ in $|z| < 1$, which is therefore a simple closed curve. Take $|w_0| < R$. Then since $|f(z)| = R$ on γ, it follows from Rouché's theorem[‡] that the functions $f(z)$ and

[†] C.A. Theorem 11, p. 107.

[‡] C.A., p. 124.

$f(z) - w_0$ have equally many zeros inside γ, N_0 say. Since $f(z)$ is not constant by the introductory hypotheses of §5, $f(z)$ assumes some values w such that $|w| < R$ inside γ, and so $N_0 \geqslant 1$. Thus $n(w) \geqslant N_0 \geqslant 1$ ($|w| < R$), and so $n(w) = 1$ for $|w| < R$.

Let now $l = l_f$ be the least upper bound of all R, if any, which satisfy the above hypotheses. If there are no such R, we put $l = 0$. Then $n(w) = 1$ for $|w| < l$. This shows that $l < +\infty$, since otherwise the inverse function $z = f^{-1}(w)$ would provide a map of the whole w plane onto $|z| < 1$, which contradicts Liouville's theorem. Also for $R > l$, we can find ϕ such that $n(R e^{i\phi}) = 0$. This remains true for $R = l$. In fact the set where $n(w) \geqslant 1$ is open, as we saw above, and so if it includes the circle $|w| = l$ it will also include $|w| = l + \epsilon$ for sufficiently small ϵ, contradicting the definition of l. This proves Lemma 5.2.

We can now prove Theorem 5.1. Let

$$\psi(z) = [f(z)]^{1/p} = a_0^{1/p} \left[1 + \frac{a_1}{p a_0} z + \dots \right].$$

Since $f(z)$ is regular and $f(z) \neq 0$ in $|z| < 1$, we can choose a single-valued branch of $\psi(z)$, which is therefore mean 1-valent by Lemma 5.1. Also $\psi(z) \neq 0$ in $|z| < 1$, and so $l = l_\psi = 0$ in Lemma 5.2. Thus for $R > 0$ we can find $w = w_R$ such that $|w_R| = R$ and $\psi(z) \neq w_R$ in $|z| < 1$.

We may therefore apply Theorem 4.17 to $\psi(z)$. (4.13) gives

$$\left| \frac{\psi'(z)}{\psi(z)} \right| \leqslant \frac{4}{1 - |z|^2},$$

and putting $z = 0$, we deduce $|a_1| \leqslant 4p |a_0|$. Also unless

$$\psi(z) = a_0^{1/p} \left(\frac{1 + z e^{i\theta}}{1 - z e^{i\theta}} \right)^2,$$

we deduce from (4.12)

$$|a_0|^{1/p} \left(\frac{1-\rho}{1+\rho} \right)^2 < |\psi(z)| < |a_0|^{1/p} \left(\frac{1+\rho}{1-\rho} \right)^2 \quad (|z| = \rho),$$

and $[(1-\rho)/(1+\rho)]^2 M(\rho, \psi)$ decreases strictly with increasing ρ ($0 < \rho < 1$). Now the inequalities of Theorem 5.1 follow on writing $f(z) = [\psi(z)]^p$. The extremal functions

$$a_0 [(1 + z e^{i\theta})/(1 - z e^{i\theta})]^{2p}$$

are mean p-valent and not zero by Lemma 5.1, since their pth roots are univalent and map $|z| < 1$ onto the plane cut along a ray from the origin to infinity.

We remarked in § 2.4 that if $f(z)$ is areally mean p-valent and hence *a fortiori* if $f(z)$ is mean p-valent in the present circumferential sense, then $f(z)$ has $q \leqslant p$ zeros. Thus the condition $f(z) \neq 0$ of Theorem 5.1 is a consequence of mean p-valency if $p < 1$.

5.2. Functions with a zero of order p at the origin. In this section we consider a function

$$f(z) = z^p + a_{p+1} z^{p+1} + \dots,$$

mean p-valent in $|z| < 1$, where p is a positive integer. In this case $f(z)$ has no zeros in $0 < |z| < 1$ and so

$$[f(z)]^{1/p} = z\left(1 + \frac{a_{p+1}}{p} z + \dots\right)$$

is single-valued and hence mean 1-valent in $|z| < 1$ and vanishes only at $z = 0$. Thus we can reduce our problem to the case $p = 1$. We show that the methods of § 1.2.1 are applicable.

THEOREM 5.2. *Let* $f(z) = z + a_2 z^2 + \dots$ *be mean 1-valent in* $|z| < 1$. *Then* $f(z)$ *belongs to the class* $\mathfrak{S}_0$ *defined in* § 1.2.1.

We suppose that $|z_0| < 1$ and that $z = \omega(\zeta)$ gives a univalent map of $|\zeta| < 1$ into $|z| < 1$ in such a way that $\omega(\zeta) \neq z_0$. Consider

$$\psi(\zeta) = f[\omega(\zeta)].$$

Then the equation $\psi(\zeta) = w$ never has more roots in $|\zeta| < 1$ than the equation $f(z) = w$ has in $|z| < 1$, since $\omega(\zeta)$ is univalent. Thus $\psi(\zeta)$ is mean 1-valent in $|\zeta| < 1$. Let $f(z_0) = R e^{i\phi_0}$. Then it follows from Lemma 5.2 that if the equation $f(z) = R e^{i\phi_0}$ has a root in $|z| < 1$ other than z_0, then we can find ϕ' such that $f(z) \neq R e^{i\phi'}$ in $|z| < 1$. In this case $\psi(\zeta) \neq R e^{i\phi'}$ in $|\zeta| < 1$. If, on the other hand, $f(z) \neq R e^{i\phi_0}$ for $z \neq z_0$, then $\psi(\zeta) \neq R e^{i\phi_0}$ in $|\zeta| < 1$.

Thus in either case we can find $\phi = \phi(R)$ such that $\psi(\zeta) \neq R e^{i\phi}$ in $|\zeta| < 1$. Now Lemma 5.2 shows that for $\rho > R$ there exists

$\phi = \phi(\rho)$ such that $\psi(\zeta) \neq \rho\, e^{i\phi(\rho)}$ in $|\zeta| < 1$. Hence Theorem 4.13 gives

$$|\psi'(0)| \leqslant 4[|\psi(0)| + R] = 4[|\psi(0)| + |f(z_0)|].$$

Hence $f(z) \in \mathfrak{S}_0$ and Theorem 5.2 is proved.

We deduce immediately†

THEOREM 5.3. *Suppose that* $f(z) = z^p + a_{p+1} z^{p+1} + \ldots$ *is mean* p-*valent in* $|z| < 1$, *where* p *is a positive integer. Then* $|a_{p+1}| \leqslant 2p$. *Further we have for* $|z| = r$ $(0 < r < 1)$

$$\frac{r^p}{(1+r)^{2p}} \leqslant |f(z)| \leqslant \frac{r^p}{(1-r)^{2p}},$$

$$|f'(z)| \leqslant \frac{p(1+r)}{r(1-r)} |f(z)| \leqslant \frac{p r^{p-1}(1+r)}{(1-r)^{2p+1}}.$$

Finally the equation $f(z) = w$ *has exactly* p *roots in* $|z| < 1$ *if* $|w| < 4^{-p}$.

THEOREM 5.4.‡ *With the hypotheses of Theorem 5.3*

$$r^{-p}(1-r)^{2p} M(r, f)$$

decreases strictly with increasing r $(0 < r < 1)$, *and so tends to* $\alpha < 1$ $(r \to 1)$, *unless* $f(z) = z^p(1 - z\,e^{i\theta})^{-2p}$. *Hence the upper bounds for* $|f(z)|, |f'(z)|$, *given in Theorem 5.3 are only attained by these functions.*

We write

$$\psi(z) = [f(z)]^{1/p} = z + \frac{a_{p+1}}{p} z^2 + \ldots.$$

By Theorem 5.2 $\psi(z) \in \mathfrak{S}_0$, and so Theorems 1.4 and 1.5 are applicable with $\psi(z)$ instead of $f(z)$. The results of Theorems 5.3 and 5.4 now follow on taking pth powers.

5.3. Regularity theorems.

Suppose that under given hypotheses a restriction results on the rate of growth of certain functions. By a regularity theorem we shall mean a theorem which states that those functions having the extremal growth

---

† In this form the result appears in Hayman[1]. The inequality $|a_{p+1}| \leqslant 2p$ has been proved for areally mean p-valent functions by Spencer[4], and the last statement of the theorem was extended to areally mean p-valent functions by Garabedian and Royden[1].

‡ For this and subsequent results of this chapter see Hayman[4].

in the class grow in a smooth or regular manner. Our first result is a regularity theorem for the maximum modulus of mean p-valent functions.

THEOREM 5.5. *Suppose that $f(z)$ is mean-p valent in $|z| < 1$. Then the limit*

$$\alpha = \lim_{r \to 1-} (1-r)^{2p} M(r, f)$$

exists finitely.

We shall need

LEMMA 5.3. *Suppose that $f(z)$ is mean p-valent and $f(z) \neq 0$ for $1 - 2\delta < |z| < 1$, where $\delta > 0$. For a fixed real θ write*

$$f(r\,e^{i\theta}) = R(r)\,e^{i\lambda(r)}.$$

Then if $1 - \delta \leqslant r_1 < r_2 < 1$ and $R = R_1, R_2$ correspond to $r = r_1, r_2$ we have

$$\log\left[R_2\left(\frac{1-r_2}{r_2+2\delta-1}\right)^{2p} \right] \leqslant \log\left[R_1\left(\frac{1-r_1}{r_1+2\delta-1}\right)^{2p} \right]$$

$$- \frac{1}{8p} \int_{r_1}^{r_2} (1-r)\,\lambda'(r)^2\,dr.$$

Consider $\qquad \phi(\zeta) = f[(1-\delta)\,e^{i\theta} + \delta\zeta].$

Then $\phi(\zeta)$ is mean p-valent and $\phi(\zeta) \neq 0$ in $|\zeta| < 1$. Thus Theorem 5.1 gives

$$\left| \frac{\phi'(\zeta)}{\phi(\zeta)} \right| = \left| \frac{\delta f'[(1-\delta)\,e^{i\theta} + \delta\zeta]}{f[(1-\delta)\,e^{i\theta} + \delta\zeta]} \right| \leqslant \frac{4p}{1-|\zeta|^2}.$$

Choose ζ so that $(1-\delta)\,e^{i\theta} + \delta\zeta = r\,e^{i\theta}$, $|\zeta| = (r+\delta-1)/\delta$. We deduce

$$\left| \frac{f'(r\,e^{i\theta})}{f(r\,e^{i\theta})} \right| \leqslant \frac{4p\delta}{\delta^2 - (r+\delta-1)^2} = \frac{4p\delta}{(r+2\delta-1)(1-r)} \quad (1-\delta < r < 1).$$

Now

$$(5.2)$$

$$e^{i\theta} \frac{f'(r\,e^{i\theta})}{f(r\,e^{i\theta})} = \frac{\partial}{\partial r} \log f(r\,e^{i\theta}) = \frac{1}{R}\frac{dR}{dr} + i\lambda'(r) = \sigma(r) + i\lambda'(r),$$

say. Then (5.2) gives

$$[\sigma(r)]^2 + [\lambda'(r)]^2 \leqslant \left[\frac{4p\delta}{(r+2\delta-1)(1-r)} \right]^2,$$

and so

$$\frac{4p\delta}{(r+2\delta-1)(1-r)} - |\sigma(r)| \geqslant \frac{[\lambda'(r)]^2}{\{4p\delta/[(r+2\delta-1)(1-r)]\} + |\sigma(r)|}$$

$$\geqslant \frac{(1-r)(r+2\delta-1)[\lambda'(r)]^2}{8p\delta}.$$

If $r \geqslant 1-\delta$, this gives

$$\sigma(r) = \frac{d\log R}{dr} \leqslant \frac{4p\delta}{(r+2\delta-1)(1-r)} - \frac{1}{8p}(1-r)[\lambda'(r)]^2.$$

On integrating from r_1 to r_2 we deduce

$$\log R_2 - \log R_1 \leqslant 2p\left[\log\left(\frac{r+2\delta-1}{1-r}\right)\right]_{r_1}^{r_2} - \frac{1}{8p}\int_{r_1}^{r_2}(1-r)[\lambda'(r)]^2\,dr.$$

This proves Lemma 5.3.

We now proceed to prove Theorem 5.5. Since $f(z)$ is mean p-valent in $|z| < 1$, $f(z)$ has $q \leqslant p$ zeros there. Thus we can find $\delta > 0$ such that $f(z) \neq 0$ in $1 - 2\delta < |z| < 1$. We then apply Lemma 5.3 and choose θ so that $M(r_2, f) = R_2$. Then $R_1 \leqslant M(r_1, f)$, and Lemma 5.3 shows that

$$\left(\frac{1-r_2}{r_2+2\delta-1}\right)^{2p} M(r_2, f)$$
$$\leqslant \left(\frac{1-r_1}{r_1+2\delta-1}\right)^{2p} M(r_1, f) \quad (1-\delta < r_1 < r_2 < 1).$$

Thus
$$\left(\frac{1-r}{r+2\delta-1}\right)^{2p} M(r, f)$$

decreases with increasing r $(1-\delta < r < 1)$, and so approaches a limit, $\alpha(2\delta)^{-2p}$ say, as $r \to 1$ from below. Hence

$$(1-r)^{2p} M(r, f) \to \alpha,$$

as required. This proves Theorem 5.5.

5.4. The radius of greatest growth.

Our aim is to prove the analogue of Theorem 5.5 for the coefficients of mean p-valent functions $f(z)$, or more generally positive powers of $f(z)$. For this purpose we now assume that

$$\alpha = \lim_{r\to 1}(1-r)^{2p} M(r, f) > 0. \tag{5.3}$$

If $\alpha = 0$, our results can be proved by more elementary methods. If (5.3) holds it follows from Theorem 2.8 that there exists a radius on which

$$\lim_{r \to 1} (1-r)^{2p} \left| f(r\, e^{i\theta_0}) \right| > 0.$$

For our (circumferentially) mean p-valent functions we can prove more than this directly.

THEOREM 5.6. *Suppose that* $f(z)$ *is mean p-valent in* $|z| < 1$ *and that* (5.3) *holds. Then there exists* θ_0 *such that* $0 \leqslant \theta_0 < 2\pi$ *and*

$$\lim_{r \to 1} (1-r)^{2p} \left| f(r\, e^{i\theta_0}) \right| = \alpha. \tag{5.4}$$

We suppose again that $f(z) \neq 0$ for $1 - 2\delta < |z| < 1$ and use Lemma 5.3. Set $r_n = 1 - 1/n$ and choose θ_n, so that $0 \leqslant \theta_n \leqslant 2\pi$ and

$$\left| f(r_n\, e^{i\theta_n}) \right| = M(r_n, f) \quad (n > 1/\delta).$$

Then it follows from (5.3) that

$$\beta_n = \left(\frac{1 - r_n}{r_n + 2\delta - 1} \right)^{2p} \left| f(r_n\, e^{i\theta_n}) \right| \to \frac{\alpha}{(2\delta)^{2p}}.$$

We take $\theta = \theta_n$ in Lemma 5.3 and obtain

$$\left(\frac{1-r}{r+2\delta-1} \right)^{2p} \left| f(r\, e^{i\theta_n}) \right| \geqslant \beta_n \quad (1-\delta \leqslant r \leqslant r_n).$$

Let θ_0 be a limit point of the sequence θ_n and let r be fixed. Then we deduce

$$\left(\frac{1-r}{r+2\delta-1} \right)^{2p} \left| f(r\, e^{i\theta_0}) \right| \geqslant \frac{\alpha}{(2\delta)^{2p}} \quad (1-\delta \leqslant r < 1).$$

Thus

$$\lim_{r \to 1} (1-r)^{2p} \left| f(r\, e^{i\theta_0}) \right| \geqslant \alpha.$$

Also by (5.3)

$$\overline{\lim_{r \to 1}} (1-r)^{2p} \left| f(r\, e^{i\theta_0}) \right| \leqslant \alpha.$$

Now Theorem 5.6 follows.

We shall also need information about $\arg f(r\, e^{i\theta_0})$. We prove

LEMMA 5.4. *With the hypotheses of Theorem 5.6 write*

$$f(r\, e^{i\theta_0}) = R(r)\, e^{i\lambda(r)}.$$

Then

$$\int_{1-\delta}^{1} (1-r)\,[\lambda'(r)]^2\,dr < +\infty.$$

Hence $\lambda(r)-\lambda(\rho)\to 0$ *and so* $\left(\dfrac{1-r}{1-\rho}\right)^{2p}\dfrac{f(r\,e^{i\theta_0})}{f(\rho\,e^{i\theta_0})}\to 1$ *uniformly as* $\rho\to 1$, *while* $\rho\leqslant r\leqslant\frac{1}{2}(1+\rho)$.

In Lemma 5.3 take $r_1=1-\delta$, $\theta=\theta_0$. Then we obtain

$$\int_{1-\delta}^{r_2} (1-r)\,[\lambda'(r)]^2\,dr \leqslant 8p\log\left[\frac{R_1(r_2+2\delta-1)^{2p}}{R_2(1-r_2)^{2p}}\right].$$

Also by Theorem 5.6 $(1-r_2)^{2p}R_2\to\alpha$ as $r_2\to 1$. Hence, making $r_2\to 1$, we deduce

$$\int_{1-\delta}^{1} (1-r)\,[\lambda'(r)]^2\,dr \leqslant 8p\log\left[\frac{R_1(2\delta)^{2p}}{\alpha}\right].$$

In particular the integral is finite.

Suppose next that a positive ϵ is given and that ρ is so near 1 that

$$\int_{\rho}^{1} (1-t)\,[\lambda'(t)]^2\,dt < \epsilon.$$

Then if $\rho\leqslant r\leqslant\frac{1}{2}(1+\rho)$ we have by Schwarz's inequality

$$|\lambda(r)-\lambda(\rho)| \leqslant \int_{\rho}^{r} |\lambda'(t)|\,dt \leqslant \left\{\int_{\rho}^{r}(1-t)\,[\lambda'(t)]^2\,dt \int_{\rho}^{r}\frac{dt}{1-t}\right\}^{\frac{1}{2}}$$

$$\leqslant \epsilon^{\frac{1}{2}}\left\{\log\frac{1-\rho}{1-r}\right\}^{\frac{1}{2}} \leqslant [\epsilon\log 2]^{\frac{1}{2}}.$$

Thus

$$\lambda(r)-\lambda(\rho) = \arg\frac{f(r\,e^{i\theta_0})}{f(\rho\,e^{i\theta_0})}\to 0,$$

as $\rho\to 1$. Since by Theorem 5.6

$$\left|\frac{f(r\,e^{i\theta_0})}{f(\rho\,e^{i\theta_0})}\right| \sim \left(\frac{1-\rho}{1-r}\right)^{2p},$$

as $r\to 1$, $\rho\to 1$ independently, the proof of Lemma 5.4 is complete.

5.5. Growth of coefficients: the case $\alpha=0$.

Suppose that $f(z)$ is mean p-valent and not zero in the annulus $1-2\delta<|z|<1$ and that $\lambda>0$. Put $\phi(z)=[f(z)]^\lambda$. Then $\phi(z)$ may be analytically continued throughout the annulus $1-2\delta<|z|<1$. Also if $\phi_2(z)$

is obtained from a branch $\phi_1(z)$ of $\phi(z)$ by analytic continuation once around the annulus in the positive direction, then

$$|\phi_2(z)/\phi_1(z)| = 1,$$

and so by the maximum modulus principle $\phi_2(z)/\phi_1(z) = e^{i\mu}$, where μ is a real constant.

Thus $\phi(z)/z^\mu$ remains one-valued in the annulus and possesses a Laurent expansion, so that

$$\phi(z) = z^\mu \sum_{n=-\infty}^{+\infty} b_n z^n. \tag{5.5}$$

We now prove

THEOREM 5.7. *Suppose that $f(z)$ is mean p-valent in $|z| < 1$, that $p\lambda > \frac{1}{4}$, and that $\phi(z) = [f(z)]^\lambda$ possesses a power series expansion* (5.5) *in an annulus $1 - 2\delta < |z| < 1$. Then*†

$$\lim_{n \to +\infty} \frac{|b_n|}{n^{2p\lambda-1}} = \frac{\alpha^\lambda}{\Gamma(2p\lambda)},$$

where α is the constant of Theorem 5.5.

We shall use the formula

$$(n+\mu)b_n = \frac{1}{2\pi i} \int_{|z|=\rho} \frac{\phi'(z)\,dz}{z^{n+\mu}} = \frac{\rho^{1-n-\mu}}{2\pi} \int_{-\pi}^{\pi} \phi'(\rho e^{i\theta}) e^{-i(n+\mu-1)\theta}\,d\theta, \tag{5.6}$$

where $1 - 2\delta < \rho < 1$. In this section we prove Theorem 5.7 when $\alpha = 0$. We choose a positive constant t such that

$$0 < t < 2, \quad t > 2 - \frac{2}{\lambda} \quad \text{and} \quad t > \frac{1}{2p\lambda}. \tag{5.7}$$

This is possible since $p\lambda > \frac{1}{4}$. Then Schwarz's inequality gives

$$\frac{1}{2\pi} \int_{-\pi}^{\pi} |\phi'(\rho e^{i\theta})|\,d\theta$$

$$\leqslant \left(\frac{1}{2\pi} \int_{-\pi}^{\pi} |\phi'(\rho e^{i\theta})|^2 |\phi(\rho e^{i\theta})|^{-t}\,d\theta\right)^{\frac{1}{2}} \left(\frac{1}{2\pi} \int_{-\pi}^{\pi} |\phi(\rho e^{i\theta})|^t\,d\theta\right)^{\frac{1}{2}}$$

$$= \lambda \left(\frac{1}{2\pi} \int_0^{2\pi} |f'(\rho e^{i\theta})|^2 |f(\rho e^{i\theta})|^{(2-t)\lambda-2}\,d\theta\right)^{\frac{1}{2}}$$

$$\times \left(\frac{1}{2\pi} \int_0^{2\pi} |f(\rho e^{i\theta})|^{\lambda t}\,d\theta\right)^{\frac{1}{2}}.$$

† Here and subsequently $\Gamma(x)$ denotes the Gamma function. Relevant properties will be found for instance in Titchmarsh [1], pp. 55–8.

Since $\alpha = 0$ we can, given $\epsilon > 0$, find $r_0(\epsilon) < 1$ such that

$$M(r, f) \leqslant \epsilon (1-r)^{-2p} \quad (r_0(\epsilon) < r < 1). \tag{5.8}$$

Since $f(z)$ is circumferentially and so *a fortiori* areally mean p-valent and (5.6) holds, we may apply Lemma 3.1 with $(2-t)\lambda$ instead of λ. Hence if $r_0(\epsilon) < r < 1$ there exists ρ such that $2r - 1 \leqslant \rho \leqslant r$ and

$$\frac{1}{2\pi} \int_0^{2\pi} |f'(\rho e^{i\theta})|^2 \, |f(\rho e^{i\theta})|^{(2-t)\lambda-2} \, d\theta \leqslant \frac{4p\epsilon^{\lambda(2-t)}}{\lambda(2-t)} (1-r)^{-2p\lambda(2-t)-1}.$$

Next we have from inequality (3.10) of Theorem 3.2 applied with λt instead of λ, and (5.8), for $0 \leqslant \rho \leqslant r$, $r_0 \leqslant r < 1$,

$$I_{\lambda t}(\rho, f)$$

$$= \frac{1}{2\pi} \int_0^{2\pi} |f(\rho e^{i\theta})|^{\lambda t} \, d\theta \leqslant M(r_0, f)^{\lambda t} + \frac{p}{2}[(\lambda t)^2 + 1] \int_{r_0}^r \frac{M(x, f)^{\lambda t} \, dx}{x}$$

$$\leqslant \epsilon^{\lambda t}(1-r_0)^{-2p\lambda t} + \frac{p}{2}[(\lambda t)^2 + 1] \int_{r_0}^r \epsilon^{\lambda t}(1-x)^{-2p\lambda t} \frac{dx}{x} < (1-r)^{1-2p\lambda t},$$

if r is sufficiently near 1. Thus if r is sufficiently near 1 we deduce

$$\frac{1}{2\pi} \int_0^{2\pi} |\phi'(\rho e^{i\theta})| \, d\theta \leqslant A(p, \lambda) \, \epsilon^{\frac{1}{2}\lambda(2-t)}(1-r)^{\frac{1}{2}[1-2p\lambda t-2p\lambda(2-t)-1]}$$

$$\leqslant A(p, \lambda) \, \epsilon^{\frac{1}{2}\lambda(2-t)}(1-r)^{-2p\lambda},$$

where ρ is some number such that $2r - 1 \leqslant \rho \leqslant r$. We choose $r = 1 - 1/n$ and apply (5.6). Then if n is sufficiently large

$$|(n+\mu) b_n| \leqslant \left(1 - \frac{2}{n}\right)^{-n-\mu} \frac{1}{2\pi} \int_0^{2\pi} |\phi'(\rho e^{i\theta})| \, d\theta$$

$$\leqslant A(p, \lambda) \, \epsilon^{\frac{1}{2}\lambda(2-t)} n^{2p\lambda}.$$

Since ϵ may be chosen as small as we please, we deduce

$$b_n = o(n^{2p\lambda-1}) \quad (n \to +\infty),$$

and this proves Theorem 5.7 when $\alpha = 0$.

5.6. The case $\alpha > 0$: the minor arc.

When $\alpha > 0$ it follows from Theorem 2.9 that θ_0 is unique in Theorem 5.6 and that $|f(r e^{i\theta})|$ is relatively small except when θ is near θ_0.

We shall deduce Theorem 5.7 from formula (5.6) in this case by obtaining an asymptotic expansion for $\phi'(\rho e^{i\theta})$ on a *major*

$arc \{\theta\colon |\,\theta-\theta_0\,| < K(1-\rho)\}$, where K is a large positive constant, and showing that the complementary *minor arc*

$$\gamma = \{\theta\colon K(1-\rho) \leqslant |\,\theta-\theta_0\,| \leqslant \pi\}$$

contributes relatively little to the integral in (5.6). It is this latter result which we prove first.

LEMMA 5.5. *Suppose that $\phi(z)$ is defined as in Theorem 5.7, that $\alpha > 0$ and that θ_0 satisfies (5.4). Then given $\eta > 0$, we can choose $K > 0$ and a positive integer n_0 with the following property. If $n > n_0$, there exists ρ in the range $1 - \dfrac{1}{n} \leqslant \rho \leqslant 1 - \dfrac{1}{2n}$ such that*

$$\frac{1}{2\pi} \int_\gamma |\,\phi'(\rho\,e^{i\theta})\,|\,d\theta < \eta n^{2p\lambda},$$

where γ is the minor arc $\{\theta\colon K(1-\rho) \leqslant |\,\theta-\theta_0\,| \leqslant \pi\}$.

We again define t to satisfy the inequalities (5.7) and have

$$\frac{1}{2\pi} \int_\gamma |\,\phi'(\rho\,e^{i\theta})\,|\,d\theta \leqslant \lambda \left(\frac{1}{2\pi} \int_\gamma |\,f'(\rho\,e^{i\theta})\,|^2 \,|\,f(\rho\,e^{i\theta})\,|^{(2-t)\lambda-2}\right)^{\frac{1}{2}}$$

$$\times \left(\frac{1}{2\pi} \int_\gamma |\,f(\rho\,e^{i\theta})\,|^{\lambda t}\,d\theta\right)^{\frac{1}{2}}. \quad (5.9)$$

Since $f(z)$ is mean p-valent, $f(z)$ satisfies (3.11) with a suitable constant $C > 0$ and $\alpha = 2p$. Thus we may apply Lemma 3.1 with $r = 1 - 1/(2n)$, $\alpha = 2p$ and $\lambda(2-t)$ instead of λ. This gives, for some ρ such that $1 - 1/n \leqslant \rho \leqslant 1 - 1/(2n)$,

$$\frac{1}{2\pi} \int_\gamma |\,f'(\rho\,e^{i\theta})\,|^2 \,|\,f(\rho\,e^{i\theta})\,|^{(2-t)\lambda-2}\,d\theta \leqslant \frac{4pC^{\lambda(2-t)}}{\lambda(2-t)} (2n)^{2p\lambda(2-t)+1}.$$

$$(5.10)$$

To estimate the second factor on the right-hand side of (5.9) we note that the number θ_0 of (5.4) certainly has the properties of θ_0 in Theorem 2.8. We may thus apply Theorem 2.9. Hence if K is chosen sufficiently large, depending on ϵ, we have for $r_0 < \rho < 1$

$$|\,f(\rho\,e^{i\theta})\,| < \frac{1}{(1-\rho)^\epsilon\,|\,\theta-\theta_0\,|^{2p-\epsilon}} \quad (K(1-\rho) \leqslant |\,\theta-\theta_0\,| \leqslant \pi).$$

Writing $|\theta - \theta_0| = x$ we deduce

$$\frac{1}{2\pi} \int_\gamma |f(\rho\,e^{i\theta})|^{\lambda t}\,d\theta \leqslant \frac{2(1-\rho)^{-\epsilon\lambda t}}{2\pi} \int_{K(1-\rho)}^\pi x^{-\lambda t(2p-\epsilon)}\,dx.$$

We choose ϵ so small that $\lambda t(2p - \epsilon) > 1$, which is possible by (5.7). Then

$$\frac{1}{2\pi} \int_\gamma |f(\rho\,e^{i\theta})|^{\lambda t}\,d\theta \leqslant \frac{(1-\rho)^{-\epsilon\lambda t}}{\pi} \int_{K(1-\rho)}^\infty x^{-\lambda t(2p-\epsilon)}\,dx$$

$$= \frac{K^{1-\lambda t(2p-\epsilon)}(1-\rho)^{1-\lambda t(2p-\epsilon)-\epsilon\lambda t}}{\pi[\lambda t(2p-\epsilon)-1]} \leqslant \frac{K^{1-\lambda t(2p-\epsilon)}}{\pi[\lambda t(2p-\epsilon)-1]}(2n)^{2p\lambda t-1}.$$
$$(5.11)$$

We note that when λ, t, C, η and ϵ have been fixed, K may then be chosen so large that

$$\frac{\lambda\,4pC^{\lambda(2-t)}}{\lambda(2-t)}\frac{K^{1-\lambda t(2p-\epsilon)}}{\pi[\lambda t(2p-\epsilon)-1]}2^{4p\lambda} < \eta^2.$$

With this choice of K Lemma 5.5 follows from (5.9), (5.10) and (5.11).

5.7. The major arc. Our next aim is to find an asymptotic formula for $f(z)$ and hence for $\phi(z)$ on the arc $|\theta - \theta_0| \leqslant K(1-\rho)$ of $|z| = \rho$, which is complementary to γ. For this purpose we define

$$r_n = 1 - \frac{1}{n}, \quad z_n = r_n\,e^{i\theta_0} \quad (n \geqslant 1),$$

where θ_0 is again the number of Theorem 5.6. We choose a fixed positive ϵ and denote by $\Delta_n = \Delta_n(\epsilon)$ the domain

$$\left\{z : \frac{\epsilon}{n} < |1 - z\,e^{-i\theta_0}| < \frac{1}{\epsilon n}, \quad |\arg(1 - z\,e^{-i\theta_0})| < \tfrac{1}{2}\pi - \epsilon\right\}.$$

Further let $$\alpha_n = n^{-2p}f(z_n),$$

and set $$f_n(z) = \frac{\alpha_n}{(1 - z\,e^{-i\theta_0})^{2p}}.$$

Thus $\alpha_n, f_n(z)$ are defined for $n \geqslant 1$ and

$$|\alpha_n| \to \alpha \quad (n \to +\infty)$$

by Theorem 5.6. We can now prove

LEMMA 5.6. *We have, as $n \to +\infty$, uniformly for z in $\Delta_n(\epsilon)$*

$$f(z) \sim f_n(z), \quad f'(z) \sim f_n'(z).$$

We shall suppose without loss in generality that $\theta_0 = 0$, since otherwise we may consider $f(z e^{i\theta_0})$, $f_n(z e^{i\theta_0})$ instead of $f(z), f_n(z)$. Then if

$$z = l_n(Z) \equiv r_n + \frac{1}{n} Z, \quad 1 - z = \frac{1}{n}(1 - Z), \quad (5.12)$$

the domain $\Delta_n(\epsilon)$ in the z plane corresponds in the Z plane to

$$\Delta_1(\epsilon) = \left\{ Z: \epsilon < |1 - Z| < \frac{1}{\epsilon}, \quad |\arg(1 - Z)| < \tfrac{1}{2}\pi - \epsilon \right\}.$$

We write
$$g_n(Z) = (1 - Z)^{2p} \frac{f[l_n(Z)]}{f(z_n)}.$$

Then for a fixed ϵ, $\Delta_n(\epsilon)$ lies in $|z| < 1$, when n is sufficiently large, and so $g_n(Z)$ is defined in $\Delta_1(\epsilon)$ for all large n. Let

$$\Delta_n' = \Delta_n(\tfrac{1}{2}\epsilon), \quad \Delta_n'' = \Delta_n(\tfrac{1}{4}\epsilon)$$

and let
$$\Delta' = \Delta_1(\tfrac{1}{2}\epsilon), \quad \Delta'' = \Delta_1(\tfrac{1}{4}\epsilon)$$

be the corresponding domains in the Z plane. Let d be the distance of Δ' from the complement of Δ''. Then the distance of Δ_n' from the complement of Δ_n'' is d/n, and hence the distance of Δ_n' from $|z| = 1$ is for large n at least d/n. Thus by Theorem 5.5

$$|f(z)| = O(n^{2p})$$

for large n, uniformly for z in Δ_n. Hence also

$$f[l_n(Z)] = O(n^{2p})$$

for Z in Δ'. This gives

$$g_n(Z) = O(1) \quad (n \to +\infty),$$

uniformly for Z in Δ'.

Suppose now that Z is real, $-1 \leqslant Z \leqslant 0$, and let $z = \rho$ correspond to Z by (5.12). Then $\rho \leqslant r_n \leqslant \tfrac{1}{2}(1 + \rho)$, and hence it follows from Lemma 5.4 that

$$g_n(Z) \sim (1 - Z)^{2p} \left(\frac{1 - r_n}{1 - \rho} \right)^{2p} = 1 \quad (n \to \infty).$$

Thus $g_n(Z)$ is uniformly bounded in Δ' for large n and $g_n(Z) \to 1$ $(n \to \infty)$ on a segment of the real axis in the interior of Δ'. Thus by Vitali's convergence theorem[†]

$$g_n(Z) \to 1, \quad g_n'(Z) \to 0 \quad (n \to \infty),$$

uniformly for Z in $\Delta_1(\epsilon)$.

Translating back to the z plane we deduce

$$\frac{n^{2p}f(z)(1-z)^{2p}}{f(z_n)} \to 1, \quad \frac{1}{n}\frac{d}{dz}\frac{n^{2p}f(z)(1-z)^{2p}}{f(z_n)} \to 0 \quad (n \to \infty),$$

uniformly for z in Δ_n. The first of these limiting relations gives

$$f(z) \sim \frac{f(z_n)}{n^{2p}}(1-z)^{-2p} = f_n(z)$$

as required. The second relation gives

$$f'(z)(1-z)^{2p} - 2p(1-z)^{2p-1}f(z) = o(n) \quad (n \to \infty).$$

Since $\epsilon < n \mid 1-z \mid < \epsilon^{-1}$ in Δ_n, we deduce

$$f'(z) = \frac{2p}{1-z}f(z) + o(n^{2p+1}),$$

and now the second asymptotic relation of Lemma 5·6 follows from the first.

5.8. Proof of Theorem 5.7.

We now conclude the proof of Theorem 5.7 when $\alpha > 0$, by proving the stronger

THEOREM 5.8. *Suppose that $\phi(z) = z^\mu \sum_{n=-\infty}^{+\infty} b_n z^n$ satisfies the hypotheses of Theorem 5.7, that $\alpha > 0$ and that θ_0 satisfies (5.4). Then*

$$nb_n \sim \frac{\phi\left[\left(1-\dfrac{1}{n}\right)e^{i\theta_0}\right]e^{-i(n+\mu)\theta_0}}{\Gamma(2p\lambda)} \quad (n \to +\infty).$$

We have by Theorem 5.6

$$\left| \phi\left[\left(1-\frac{1}{n}\right)e^{i\theta_0}\right] \right| = \left| f\left[\left(1-\frac{1}{n}\right)e^{i\theta_0}\right] \right|^\lambda \sim \alpha^\lambda n^{2p\lambda} \quad (n \to +\infty).$$

† Titchmarsh[1], p. 168.

Thus Theorem 5.7 follows at once from Theorem 5.8. The latter result is significantly stronger, since it gives information about $\arg b_n$ as well as $|b_n|$.

We suppose, as we may without loss in generality, that $\theta_0 = 0$, since otherwise we may consider $f(z e^{i\theta_0})$, $\phi(z e^{i\theta_0})$ instead of $f(z)$, $\phi(z)$. Write

$$(1-z)^{-2p\lambda} = \sum_{n=0}^{\infty} c_n z^n.$$

Then

$$c_n = \frac{2p\lambda(2p\lambda+1)\dots(2p\lambda+n-1)}{1.2\dots.n}$$
$$= \frac{\Gamma(n+2p\lambda)}{\Gamma(n+1)\,\Gamma(2\lambda p)} \sim \frac{n^{2p\lambda-1}}{\Gamma(2p\lambda)} \quad (n\to+\infty).$$

We also set $\phi_n(z) = f_n(z)^{\lambda}$, so that

$$\phi_n(z) = \alpha_n^{\lambda} \sum_{m=0}^{\infty} c_m z^m,$$

and hence

$$n c_n \alpha_n^{\lambda} = \frac{1}{2\pi i} \int_{|z|=\rho} \frac{\phi_n'(z)\,dz}{z^n} = \frac{\rho^{1-n}}{2\pi} \int_{-\pi}^{\pi} \phi_n'(\rho e^{i\theta}) e^{-i(n-1)\theta}\,d\theta \quad (0<\rho<1).$$

We also recall (5.6):

$$(n+\mu)\,b_n = \frac{\rho^{1-n-\mu}}{2\pi} \int_{-\pi}^{+\pi} \phi'(\rho e^{i\theta}) e^{-i(n+\mu-1)\theta}\,d\theta \quad (1-2\delta<\rho<1).$$

Thus

$$2\pi\rho^{n-1}[(n+\mu)\,\rho^{\mu}b_n - nc_n\alpha_n^{\lambda}]$$
$$= \int_{-\pi}^{\pi} [\phi'(\rho e^{i\theta}) e^{-i\mu\theta} - \phi_n'(\rho e^{i\theta})] e^{-i(n-1)\theta}\,d\theta. \quad (5.13)$$

We now suppose $\eta>0$ and choose K so large, that for $n>n_0$ we can find ρ in the range $1-\dfrac{1}{n} \leqslant \rho \leqslant 1-\dfrac{1}{2n}$ such that

$$\int_{\gamma} |\phi'(\rho e^{i\theta})|\,d\theta < \eta n^{2p\lambda},$$

where γ denotes the arc $K(1-\rho) \leqslant |\theta| \leqslant \pi$. This is possible by Lemma 5.5. By applying that lemma to $(1-z)^{-2p\lambda}$ instead of $\phi(z)$ we may also suppose that

$$\int_{\gamma} |\phi_n'(\rho e^{i\theta})|\,d\theta < \eta n^{2p\lambda},$$

since $|\alpha_n|$ remains bounded as $n \to +\infty$. Thus

$$\left| \int_\gamma [\phi'(\rho\, e^{i\theta})\, e^{-i\mu\theta} - \phi_n'(\rho\, e^{i\theta})]\, e^{-i(n-1)\theta}\, d\theta \right|$$

$$\leqslant \int_\gamma [|\phi'(\rho\, e^{i\theta})| + |\phi_n'(\rho\, e^{i\theta})|]\, d\theta < 2\eta n^{2p\lambda}. \quad (5.14)$$

If γ' denotes the complementary arc $0 \leqslant |\theta| \leqslant K(1-\rho)$, then γ' lies in $\Delta_n(\epsilon)$ for all large n if ϵ is a sufficiently small fixed number depending on K. For we have on γ'

$$\frac{1}{2n} \leqslant |1 - z| \leqslant 1 - \rho + |e^{i\theta} - 1| \leqslant (1+K)(1-\rho) \leqslant \frac{1+K}{n},$$

and

$$|\arg(1-z)| \leqslant \tan^{-1} \frac{\rho \sin\theta}{1 - \rho\cos\theta} \leqslant \tan^{-1} \frac{\theta}{1-\rho} \leqslant \tan^{-1} K.$$

Thus Lemma 5.6 gives, with $z = \rho\, e^{i\theta}$,

$$\phi'(z) = \lambda f'(z)\, [f(z)]^{\lambda-1} \sim \lambda f_n'(z)\, [f_n(z)]^{\lambda-1} = \phi_n'(z),$$

as $n \to \infty$, uniformly on γ'. Since $\theta = o(1)$ on γ' we deduce

$$\phi'(\rho\, e^{i\theta})\, e^{-i\mu\theta} \sim \phi'(\rho\, e^{i\theta}) \sim \phi_n'(\rho\, e^{i\theta}),$$

and so

$$\phi'(\rho\, e^{i\theta})\, e^{-i\mu\theta} - \phi_n'(\rho\, e^{i\theta}) = o[\phi_n'(\rho\, e^{i\theta})] = o(1-\rho)^{-2p\lambda-1} = o(n^{2p\lambda+1})$$

as $n \to +\infty$, uniformly on γ'. Thus

$$\left| \int_{\gamma'} [\phi'(\rho\, e^{i\theta})\, e^{-i\mu\theta} - \phi_n'(\rho\, e^{i\theta})]\, e^{-i(n-1)\theta}\, d\theta \right|$$

$$\leqslant \int_{\gamma'} |\phi'(\rho\, e^{i\theta})\, e^{-i\mu\theta} - \phi_n'(\rho\, e^{i\theta})|\, d\theta$$

$$= o(n^{2p\lambda+1})\, 2K(1-\rho) = o(n^{2p\lambda}). \quad (5.15)$$

Since γ, γ' together make up the whole range $-\pi \leqslant \theta \leqslant \pi$, we deduce from (5.13), (5.14) and (5.15)

$$|2\pi\rho^{n-1}[(n+\mu)\rho^\mu b_n - nc_n\alpha_n^\lambda]| < [2\eta + o(1)]\, n^{2p\lambda}$$

for large n. Also

$$\rho^{n-1} \geqslant \left(1 - \frac{1}{n}\right)^n \geqslant \frac{1}{4} \quad (n \geqslant 2),$$

η is arbitrarily small, and $\rho^\mu \to 1$ as $n \to +\infty$. Thus

$$b_n = \frac{n c_n \alpha_n^\lambda}{(n+\mu)\rho^\mu} + o(n^{2p\lambda-1})$$

$$\sim \frac{n^{2p\lambda-1}\alpha_n^\lambda}{\Gamma(2p\lambda)} = \frac{[f(z_n)]^\lambda}{n\Gamma(2p\lambda)} = \frac{\phi(z_n)}{n\Gamma(2p\lambda)} \quad (n \to +\infty),$$

where $z_n = 1 - 1/n$. This proves Theorem 5.8.

5.9. Applications: the case $\lambda = 1$. The case $\lambda = 1$ of Theorem 5.7 is of most interest, since it refers directly to the coefficients of mean p-valent functions. If

$$f(z) = \sum_0^\infty a_n z^n$$

is mean p-valent in $|z| < 1$ and $p > \frac{1}{4}$, then

$$\frac{|a_n|}{n^{2p-1}} \to \frac{\alpha}{\Gamma(2p)} \quad (n \to \infty),$$

where α is the constant of Theorem 5.5. Whenever we can obtain bounds for α, we obtain correspondingly sharp bounds for the asymptotic growth of the coefficients. Thus we have for instance

THEOREM 5.9. *Suppose that p is a positive integer and that*

$$f(z) = z^p + \sum_{n=p+1}^\infty a_n z^n$$

is mean p-valent in $|z| < 1$. Then

$$\lim_{n \to \infty} \frac{|a_n|}{n^{2p-1}} = \frac{\alpha}{(2p-1)!},$$

where $\alpha < 1$, except when $f(z) = z^p(1 - z\,e^{i\theta})^{-2p}$.

In fact the limiting relation holds by Theorem 5.7 and

$$\alpha = \lim_{r \to 1-} (1-r)^{2p} M(r, f).$$

Also, by Theorem 5.4, $\alpha < 1$ except for the functions $z^p(1 - z\,e^{i\theta})^{2p}$.

The case $p = 1$ is of particular interest. Theorem 5.9 shows that in this case

$$\lim_{n \to \infty} \frac{|a_n|}{n} = \alpha < 1,$$

unless $f(z) = z + 2z^2 e^{i\theta} + 3z^3 e^{2i\theta} + \dots$. Thus in any case we have

$$|a_n| \leqslant n \quad \text{for} \quad n > n_0(f).$$

This gives Bieberbach's conjecture, which was referred to in Chapter 1, for a fixed function $f(z)$ and all sufficiently large n.

We have similarly

THEOREM 5.10. *Suppose that* $f(z) = \sum\limits_{n=0}^{\infty} a_n z^n$ *is mean* p-*valent and* $f(z) \neq 0$ *in* $|z| < 1$, *where* $p > \frac{1}{4}$. *Then*

$$\lim_{n \to \infty} \frac{|a_n|}{n^{2p-1}} = \frac{\alpha}{\Gamma(2p)},$$

where $\alpha \leqslant |a_0| \, 4^p$ *with equality only for the functions* $a_0 \left(\dfrac{1 + z e^{i\theta}}{1 - z e^{i\theta}} \right)^{2p}$.

The limiting relation is again a consequence of Theorem 5.7. The inequality for α follows from the last statement of Theorem 5.1.

5.10. Functions with k-fold symmetry.

Suppose now more generally that k is a positive integer and that

$$f_k(z) = a_k z^k + a_{2k} z^{2k} + \dots + a_{nk} z^{nk} + \dots$$

is mean p-valent in $|z| < 1$. Then

$$\phi(z) = f_k(z^{1/k}) = a_k z + a_{2k} z^2 + \dots$$

is mean (p/k)-valent there and conversely. For to each root z_0 of the equation $\phi(z) = w$ there correspond exactly k roots of the equation $f_k(z) = w$, given by $z^k = z_0$. Thus for any w

$$n[w, f_k(z)] = kn[w, \phi(z)],$$

and so for any positive R

$$p[R, \phi] = \frac{1}{k} p[R, f_k].$$

More generally if ν is a positive integer and

$$f_k(z) = z^\nu + a_{k+\nu} z^{k+\nu} + a_{2k+\nu} z^{2k+\nu} + \dots \tag{5.16}$$

is mean p-valent in $|z| < 1$, then

$$[f_k(z)]^k = z^{\nu k}(1 + a_{k+\nu} z^k + \dots)^k = z^{\nu k} + b_{\nu+1} z^{(\nu+1)k} + \dots$$

8

is mean (pk)-valent there by Lemma 5.1, and so by the above remark

$$f(z) = [f_k(z^{1/k})]^k = z^\nu + \sum_{n=\nu+1}^{\infty} b_n z^n$$

is mean p-valent in $|z| < 1$. Thus

$$\alpha = \lim_{r \to 1} (1-r)^{2p} M[r, f(z)] \qquad (5.17)$$

exists. Further

$$f_k(z^{1/k}) = [f(z)]^{1/k} = z^{\nu/k}(1 + a_{k+\nu} z + a_{2k+\nu} z^2 + \dots).$$

Thus we may apply Theorem 5.7 to $f(z)$ with $\lambda = 1/k$. We deduce

THEOREM 5.11. *If $f_k(z)$, given by (5.16), is mean p-valent in* $|z| < 1$, *where $1 \leqslant k < 4p$, then*

$$\lim_{n \to \infty} \frac{|a_{nk+\nu}|}{n^{2p/k-1}} = \frac{\alpha^{1/k}}{\Gamma(2p/k)},$$

where α is given by (5.17).

If conversely $f(z) = z^\nu + \dots$ is any function mean p-valent in $|z| < 1$, then $f_k(z) = [f(z^k)]^{1/k}$ is mean p-valent in $|z| < 1$ and of the form (5.16), provided that $f_k(z)$ is regular, i.e. provided that all the zeros of $f(z)$ at points other than the origin have multiplicities that are multiples of k.

We note in particular the case $\nu = p$. In this case $f(z)$ and $f_k(z)$ each have a zero of order p at the origin and so they have no other zeros. To each $f(z)$ there corresponds an $f_k(z)$ and conversely. We have

THEOREM 5.12. *Suppose that*

$$f_k(z) = z^p + a_{p+k} z^{p+k} + a_{p+2k} z^{p+2k} + \dots$$

is mean p-valent in $|z| < 1$ and that $1 \leqslant k < 4p$. Then

$$\lim_{n \to \infty} \frac{|a_{p+nk}|}{n^{2p/k-1}} = \frac{\alpha^{1/k}}{\Gamma(2p/k)},$$

and $\alpha \leqslant 1$, with equality only for the mean p-valent functions

$$f_k(z) = z^p(1 - z^k e^{i\theta})^{-2p/k}.$$

In fact in this case

$$f(z) = [f_k(z^{1/k})]^k = z^p + \dots$$

is mean p-valent in $|z| < 1$, and so by Theorem 5.4

$$\alpha = \lim_{r \to 1} (1-r)^{2p} M(r, f) \leqslant 1,$$

with equality only for the functions $f(z) = z^p(1 - z e^{i\theta})^{-2p}$ which correspond to $f_k(z) = z^p(1 - z^k e^{i\theta})^{-2p/k}$ as required.

We note the special case of an odd univalent function

$$f_2(z) = z + a_3 z^3 + a_5 z^5 + \dots$$

Taking $p = 1$, $k = 2$, we deduce from Theorem 5.12 that

$$\lim_{n \to \infty} |a_{2n+1}| < 1$$

in this case, except when $f_2(z) = z + z^3 e^{i\theta} + z^5 e^{2i\theta} + \dots$. In particular,

$$|a_{2n+1}| \leqslant 1 \quad (n > n_0(f)).$$

Nevertheless, for any fixed $n \geqslant 2$ an odd univalent function $f_2(z)$ can be found† for which $|a_{2n+1}| > 1$. This example reduces to some extent the value of the evidence in favour of the Bieberbach conjecture that is given by Theorem 5.9.

5.11. Some further results.

Some inequalities for the class $\mathfrak{F}$ of functions $\quad f(z) = z + a_2 z^2 + \dots$

circumferentially mean 1-valent in $|z| < 1$, which go beyond Theorem 5.3 have recently been proved by Jenkins [1, 3]. He obtained the exact upper bound for $|f(r)|$, when $0 < r < 1$ and either $|a_2|$ or $|f(-\rho)|$ is given, where ρ is fixed and $0 < \rho < 1$. In either case the extremals are univalent functions with real coefficients. It follows from Theorem 1.10 that for these extremals $|a_n| \leqslant n$ and so

$$|f(\rho)| + |f(-\rho)| = f(\rho) - f(-\rho) = 2(\rho + a_3 \rho^3 + a_5 \rho^5 + \dots)$$

$$\leqslant 2 \sum_{n=0}^{\infty} (2n+1) \rho^{2n+1},$$

$$|f(\rho)| + |f(-\rho)| \leqslant \frac{\rho}{(1+\rho)^2} + \frac{\rho}{(1-\rho)^2}.$$

This latter inequality remains valid for all $f(z) \in \mathfrak{F}$, and hence $|a_3| \leqslant 3$ for $f(z) \in \mathfrak{F}$. An unpublished example of Spencer shows

† Schaeffer and Spencer[1]. Earlier Fekete and Szegö[1] showed that the exact upper bound for $|a_5|$ in this case is $\frac{1}{2} + e^{-\frac{2}{3}} = 1 \cdot 01 \dots$

this to be false in general for areally mean 1-valent functions, though $|a_2| \leqslant 2$ remains true. We shall prove $|a_3| \leqslant 3$ for univalent $f(z)$ by Löwner's original method in the next chapter.

Jenkins's results also imply that if $|a_2|$ is given and $f(z) \in \mathfrak{F}$, then

$$\alpha = \lim_{r \to 1-} (1-r)^2 M(r, f) \leqslant \psi(|a_2|) = 4b^{-2} \exp(2 - 4b^{-1}),$$

where $b = 2 - (2 - |a_2|)^{\frac{1}{2}}$, and this inequality is sharp for $0 \leqslant |a_2| \leqslant 2$.

If $f(z) = z^p + a_{p+1}z^{p+1} + \dots$ is mean p-valent in $|z| < 1$, then $[f(z)]^{1/p} \in \mathfrak{F}$, and from this Jenkins deduced the sharp inequality $|a_{p+2}| \leqslant p(2p+1)$ in this case. It also follows that the number α in Theorem 5.12 satisfies the inequality

$$\alpha \leqslant \psi\left(\frac{k|a_{p+k}|}{p}\right),$$

which is again sharp for given $|a_{p+k}|$. These results are outside the scope of this book and will appear in a forthcoming tract by Jenkins in the Ergebnisse series.

Regularity theorems for the means $I_\lambda(r, f)$ and their derivatives, similar to Theorems 5.5 and 5.7, together with a discussion of the behaviour of $\arg a_n$, and a different (but slightly longer) treatment of the minor arc which does not depend on the deep Theorem 2.9, will be found in Hayman [4].

CHAPTER 6

THE LÖWNER THEORY

6.0. Introduction. We shall in this chapter give a deeper theory due to Löwner [2], which enables us to obtain sharp bounds for the class $\mathfrak{S}$ of functions

$$f(z) = z + a_2 z^2 + \ldots$$

univalent in $|z| < 1$, and which do not seem to be accessible by the methods of Chapter 1. We shall say that the class $\mathfrak{S}'$ is dense in $\mathfrak{S}$, if $\mathfrak{S}'$ is a subclass of $\mathfrak{S}$ and if every function $f(z) \in \mathfrak{S}$ can be approximated by a sequence of functions $f_n(z) \in \mathfrak{S}'$ so that $f_n(z) \to f(z)$ uniformly on every compact subset of $|z| < 1$ as $n \to \infty$. It will then follow that pth derivatives at an arbitrary point in $|z| < 1$, and in particular the coefficients of $f_n(z)$, approach those of $f(z)$ as $n \to \infty$. Thus bounds obtained for the class $\mathfrak{S}'$ will remain true for the wider class $\mathfrak{S}$.

Löwner's basic result can now be stated as follows:

THEOREM 6.1. *Suppose $t_0 > 0$ and let $\kappa(t)$ be a continuous complex-valued function of t in the range $0 \leqslant t \leqslant t_0$, satisfying $|\kappa(t)| = 1$. Let $w = f(z, t)$ be the solution of the differential equation*

$$\frac{\partial w}{\partial t} = -w \frac{1 + \kappa(t) w}{1 - \kappa(t) w} \quad (0 \leqslant t \leqslant t_0), \tag{6.1}$$

such that $f(z, 0) = z$. Then the functions $e^{t_0} f(z, t_0)$ for varying t_0 and functions $\kappa(t)$ form a dense subclass of $\mathfrak{S}$.

We shall prove this result in the first part of the chapter by showing (a) that if $w = f(z, t) = e^{-t}(z + a_2(t) z^2 + \ldots)$ maps $|z| < 1$ onto $|w| < 1$ cut along a sectionally analytic slit, then the functions $e^t f(z, t)$ are dense in $\mathfrak{S}$, (b) that these functions are in fact the solutions of a differential equation (6.1) with a suitable function $\kappa(t)$, and (c) that given any continuous function $\kappa(t)$ in Theorem 6.1, then the corresponding solution $f(z, t)$ of (6.1) exists uniquely and $e^t f(z, t) \in \mathfrak{S}$.

The proof of Theorem 6.1 is long and not easy. It is, however, fully justified by its many beautiful and simple applications. Let $f(z) = z + a_2 z^2 + \ldots \in \mathfrak{S}$. In the last part of the chapter we shall give some of these applications including the exact upper bound for $|a_3|$ and all the coefficients of the inverse function $z = f^{-1}(w)$, as well as the bounds for the arguments of $f(z)/z$ and $f'(z)$ and the radii of convexity and starlikeness.

6.1. Boundary behaviour in conformal mapping. In this section we prove two preliminary results:

LEMMA 6.1. *Suppose that $w = \phi(z)$ maps a domain D_1 in $|z| < 1$ $(1, 1)$ conformally onto a domain D_2 lying in $|w| < 1$. For any point z_0 let $l(R)$ be the total length of the image in D_2 of that part of the circumference $|z - z_0| = R$ which lies in D_1. Then if $R_1 > 0$, $k > 1$, there exists R such that $R_1 < R < kR_1$ and $l(R) \leqslant \pi (2/\log k)^{\frac{1}{2}}$. In particular, there exists a sequence R_n, decreasing to zero as $n \to \infty$, such that $l(R_n) \to 0$ as $n \to \infty$.*

We consider the mapping $z = \phi^{-1}(w)$ of D_2 onto D_1 and put

$$\psi(w) = \phi^{-1}(w) - z_0.$$

Then the level curves γ_R in D_2 corresponding to $|z - z_0| = R$ are the level curves $|\psi(w)| = R$. Let $l(R)$ be their total length. Then since $\psi(w)$ is univalent, we may apply Theorem 2.1 with $p(R) \leqslant 1$, $A \leqslant \pi$. This gives

$$\int_{R_1}^{kR_1} \frac{l(R)^2 dR}{R} \leqslant \int_0^\infty \frac{l(R)^2 dR}{Rp(R)} \leqslant 2\pi A \leqslant 2\pi^2.$$

If l is the lower bound of $l(R)$ in $R_1 < R < kR_1$, we deduce

$$l^2 \log k \leqslant 2\pi^2,$$

$$l \leqslant \pi \left(\frac{2}{\log k} \right)^{\frac{1}{2}},$$

as required.

If we now define R_n' inductively by $R_0' = 1$, $R_{n+1}' = e^{-n} R_n'$, then it follows that there exists R_n such that $R_{n+1}' < R_n < R_n'$ and

$$l(R_n) < \pi \Big/ \sqrt{\left(\frac{2}{n} \right)}.$$

This completes the proof of Lemma 6.1.

We shall also need

LEMMA 6.2. *Let γ be a simple arc of length $l < 1$ which lies in* $|z| < 1$ *but approaches* $|z| = 1$ *at both ends. Let D be the set of all those points P in* $|z| < 1$ *such that any curve joining the origin to P in* $|z| < 1$ *meets γ. Then the diameter of D is at most l.*†

We may parametrize the curve γ by $z = \alpha(t)$ $(a < t < b)$. Let t_n $(n = 0, 1, 2, ...)$ be an increasing sequence of numbers such that

$$a < t_n < b \quad (n \geqslant 1),$$
$$t_n \to b \quad (n \to \infty).$$

Then $\qquad \sum_{r=1}^{n} |\alpha(t_r) - \alpha(t_{r-1})| \leqslant l,$

and so $\sum_{r=1}^{\infty} |\alpha(t_r) - \alpha(t_{r-1})| \leqslant l,$ $\sum_{r=1}^{\infty} [\alpha(t_r) - \alpha(t_{r-1})]$ converges.

Thus $\alpha(t_r)$ approaches a definite limit $\alpha(b)$ as $r \to \infty$ and $\alpha(b)$ is clearly independent of the sequence t_r. Thus $\alpha(t) \to \alpha(b)$ as $t \to b-$ and similarly $\alpha(t)$ approaches a finite limit $\alpha(a)$ as $t \to a+$.

We write $z_1 = \alpha(a)$, $z_2 = \alpha(b)$. Then $|z_1| = |z_2| = 1$ by hypothesis. Let $z_3 = \frac{1}{2}(z_1 + z_2)$. If z is any point on γ, then we have

$$|z - z_1| + |z - z_2| \leqslant l,$$

since γ has length l and so z lies inside an ellipse of centre z_3 and major axis l. Hence z lies inside the circle C of centre z_3 and radius $\frac{1}{2}l$. This circle has diameter l, and since C contains z_1 on $|z| = 1$, C cannot contain the origin. Any other point P in $|z| < 1$ but outside C can clearly be joined to the origin by a curve lying in $|z| < 1$ and outside C, and so P lies outside D. Thus D lies inside C and so has diameter at most l.

6.2. Transformations. Let

$$f(z) = \beta(z + a_2 z^2 + ...), \tag{6.2}$$

where $\beta > 0$, be univalent in $|z| < 1$ and satisfy $|f(z)| < 1$. Such a function $f(z)$ will be called a *transformation*. The transformation $w = f(z)$ maps $|z| < 1$ onto a domain D in $|w| < 1$. We denote by $S = S_f$ the set of all points of $|w| < 1$ not in D, and by $d = d_f$ the diameter of S_f. We ignore the trivial case when S_f is null and $f(z) = z$.

† The diameter of a point set E is the upper bound of distances $|z_1 - z_2|$ of pairs of points z_1, z_2 in E.

We shall say that two points z, w on $|z|=1$ and the frontier of D respectively *correspond* by the transformation $w=f(z)$, if there exists a sequence z_n, such that

$$|z_n| < 1 \quad (n \geqslant 1)$$

and
$$z_n \to z, \quad f(z_n) \to w \quad (n \to \infty).$$

Let $B = B_f$ be the set of all points of $|z|=1$, which correspond to points of S. We note that points of $|z|=1$ not in B can correspond only to points on $|w|=1$. We write $\delta = \delta_f$ for the diameter of B.

Our aim is to study the limiting behaviour of transformations when δ or d is small. In this case D approximates to $|w| < 1$ and $f(z)$ approximates to z. Our first aim is to show that, if either of δ or d is small, then so is the other and in this case β is nearly equal to 1.

LEMMA 6.3. *If $f(z)$ is a transformation given by* (6.2), *then we have with the above notation* $1 - d \leqslant \beta \leqslant 1$.

The inequality $\beta \leqslant 1$ follows from Schwarz's lemma. To prove $\beta \geqslant 1 - d$, we may suppose $d < 1$. It follows from Lemma 5.2 that if S meets $|w| = r$ for some $r < 1$, then S meets $|w| = \rho$ for $r < \rho < 1$. It follows that S has at least one limit point $e^{i\theta}$ on $|w| = 1$. Thus S lies entirely in $|w - e^{i\theta}| \leqslant d$, and so in $|w| \geqslant 1 - d$. Thus D contains the circle $|w| < 1 - d$.

Hence the inverse function $z = f^{-1}(w)$ maps $|w| < 1 - d$ onto a subdomain of $|z| < 1$ and

$$f^{-1}[(1-d)w] = \frac{(1-d)w}{\beta} + \dots$$

satisfies the conditions of Schwarz's lemma. Thus $1 - d \leqslant \beta$ as required and Lemma 6.3 is proved.

LEMMA 6.4. *We have with the above notation*

$$\delta \leqslant 4\pi[\log(2/d)]^{-\frac{1}{2}}, \quad d \leqslant 4\pi[\log(2/\delta)]^{-\frac{1}{2}}.$$

To prove the first inequality, we may assume $d < 2e^{-4\pi^2}$. For since $\delta \leqslant 2$, the inequality is trivial otherwise.

Let w_0 be a limit point of S_f on $|w| = 1$. Then S_f lies entirely in $|w - w_0| \leqslant d$, and so for $R > d$ that arc, c_R say, of $|w - w_0| = R$

which lies in $|w| < 1$, lies also in D. Let γ_R be the image of c_R by $z = f^{-1}(w)$, and let $l(R)$ be the length of γ_R. Then by Lemma 6.1, with $R_1 = d$, $k = d^{-1}$, we can choose R_0 satisfying

$$d < R_0 < 1, \quad l(R_0) \leqslant \pi \left(\tfrac{1}{2} \log \frac{1}{d} \right)^{-\frac{1}{2}} \leqslant \pi \left(\tfrac{1}{4} \log \frac{2}{d} \right)^{-\frac{1}{2}} < 1.$$

Next since c_{R_0} separates $w = 0$ from S_f, it follows that γ_{R_0} separates $z = 0$ from B_f. Hence by Lemma 6.2

$$\delta_f \leqslant l(R_0) \leqslant 2\pi \left(\log \frac{2}{d} \right)^{-\frac{1}{2}},$$

as required. The proof of the second inequality of Lemma 6.4 is similar, and so the lemma is proved.

6.2.1. We shall also need the following:

LEMMA 6.5. *Suppose that $\psi(z) = u + iv$ is regular in $|z| < 1$, that $u(z)$ has constant sign there and $v(0) = 0$. Suppose further that*

$$u(r\, e^{i\phi}) \to 0 \quad (r \to 1),$$

uniformly for $\delta \leqslant |\phi - \phi_0| \leqslant \pi$. Then we have

$$\psi(z) = \psi(0) \left[\frac{e^{i\phi_0} + z}{e^{i\phi_0} - z} + \epsilon(z) \right],$$

where $|\epsilon(z)| < 5\delta \, |e^{i\phi_0} - z|^{-2}$ for $|e^{i\phi_0} - z| > 2\delta$, $|z| < 1$.

We may without loss in generality assume $\phi_0 = 0$. Then we have by Poisson's formula for $|z| < r < 1$,

$$\psi(z) = \frac{1}{2\pi} \int_{-\pi}^{+\pi} u(r\, e^{i\theta}) \frac{r\, e^{i\phi} + z}{r\, e^{i\phi} - z} \, d\phi + iC,$$

where C is a constant in $|z| < r$, and since $\psi(0)$ is real, $C = 0$. Thus we have for a fixed z,

$$\psi(z) = \frac{1}{2\pi} \int_{-\delta}^{\delta} u(r\, e^{i\phi}) \frac{r\, e^{i\phi} + z}{r\, e^{i\phi} - z} \, d\phi + o(1) \quad (r \to 1). \qquad (6.3)$$

Writing $\psi(0) = \alpha$ we obtain

$$\frac{1}{2\pi} \int_{-\delta}^{\delta} u(r\, e^{i\phi}) \, d\phi \to \alpha \quad (r \to 1). \qquad (6.4)$$

Also

$$\frac{r\, e^{i\phi} + z}{r\, e^{i\phi} - z} \to \frac{e^{i\phi} + z}{e^{i\phi} - z},$$

and if $|z-1| \geqslant 2\delta$, $|\phi| \leqslant \delta$ we have $|e^{i\phi}-1| \leqslant \delta$ and so

$$\left| \frac{e^{i\phi}+z}{e^{i\phi}-z} - \frac{1+z}{1-z} \right| = \left| \frac{2z(e^{i\phi}-1)}{(e^{i\phi}-z)(1-z)} \right|$$

$$\leqslant \frac{2\delta}{|1-z|[|1-z|-\delta]} \leqslant \frac{4\delta}{|1-z|^2}.$$

Thus if $|z-1| > 2\delta$, $|\phi| < \delta$ and r is sufficiently near to 1 then

$$\left| \frac{r e^{i\phi}+z}{r e^{i\phi}-z} - \frac{1+z}{1-z} \right| < \frac{5\delta}{|1-z|^2}.$$

Hence we deduce from (6.3), making $r \to 1$,

$$\left| \psi(z) - \frac{1+z}{1-z} \frac{1}{2\pi} \int_{-\delta}^{\delta} u(r e^{i\phi}) \, d\phi \right|$$

$$\leqslant \frac{5\delta}{|1-z|^2} \frac{1}{2\pi} \int_{-\delta}^{\delta} |u(r e^{i\phi})| \, d\phi + o(1).$$

Using (6.4) and the fact that u has constant sign in $|z| < 1$ we deduce

$$\left| \psi(z) - \alpha \frac{1+z}{1-z} \right| \leqslant \frac{5\delta|\alpha|}{|1-z|^2},$$

if $|1-z| \geqslant 2\delta$, and this is Lemma 6.5.

6.3. Structure of infinitesimal transformations. We are now able to describe infinitesimal transformations, i.e. those for which $\delta(f)$ and $d(f)$ are small.

LEMMA 6.6. *Let $f_n(z) = \beta_n(z + \ldots)$ be a sequence of transformations of $|z| < 1$, let $d(f_n)$, $\delta(f_n)$ be defined as in § 6.2 and suppose that either $d(f_n)$ or $\delta(f_n) \to 0$ as $n \to \infty$. Then*

$$f_n(z) \to z \quad (n \to \infty), \tag{6.5}$$

uniformly in $|z| < 1$ and in particular $\beta_n \to 1$.

Further, if z_n is a point on $|z| = 1$ which corresponds to a point w_n in $|w| < 1$ by $w = f_n(z)$, and if $0 < r < 1$, then

$$f_n(z) - z \sim -(1-\beta_n) z \frac{z_n+z}{z_n-z} \quad (n \to \infty), \tag{6.6}$$

uniformly in $|z| \leqslant r$.

It follows from Lemma 6.4 that if $\delta(f_n) \to 0$ then $d(f_n) \to 0$ and conversely. It then follows from Lemma 6.3 that $\beta_n \to 1$. Write

$$\psi_n(z) = \log \frac{f_n(z)}{z} = u_n(z) + iv_n(z).$$

Since $f_n(z)$ vanishes only at $z = 0$, $\psi_n(z)$ is regular in $|z| < 1$, and since $f_n(z)$ satisfies the hypotheses of Schwarz's lemma $u_n(z) \leqslant 0$ in $|z| < 1$. Let $z_n = e^{i\phi_n}$ correspond to w_n, where $|w_n| < 1$, by $w = f_n(z)$. Then if $|e^{i\phi} - e^{i\phi_n}| > \delta(f_n)$, the point $z = e^{i\phi}$ corresponds only to points w on $|w| = 1$, and given $\delta > 0$, this will be true for $\delta \leqslant |\phi - \phi_n| \leqslant \pi$ if $n > n_0(\delta)$, since $\delta(f_n) \to 0$. In this case

$$u_n(r\,e^{i\phi}) \to 0 \quad (r \to 1),$$

uniformly for $\delta \leqslant |\phi - \phi_n| \leqslant \pi$ and a fixed $n > n_0(\delta)$.

We now apply Lemma 6.5 and obtain

$$\psi_n(z) = \log \beta_n \left[\frac{z_n + z}{z_n - z} + \epsilon_n(z) \right], \tag{6.7}$$

where $|\epsilon_n(z)| < 5\delta |z - z_n|^{-2}$ for $|z - z_n| > 2\delta$. Thus if η is fixed and positive we may choose n_1 so large that

$$|\psi_n(z)| \leqslant \eta$$

if $n > n_1$ and $|z - z_n| \geqslant \eta$, and then

$$|f_n(z) - z| = |z| \, |e^{\psi_n(z)} - 1| \leqslant e^\eta - 1 < 2\eta \tag{6.8}$$

if $\eta < \frac{1}{2}$. We also suppose n so large that the end-points of the arc $|z - z_n| = \eta$ on $|z| = 1$ correspond to points on $|w| = 1$ only. By (6.8) we have

$$|f_n(z) - z| \leqslant 2\eta \quad \text{and so} \quad |f_n(z) - z_n| \leqslant 3\eta$$

on this arc. The values which $w = f_n(z)$ assumes for $|z - z_n| < \eta$ form a domain Δ in $|w| < 1$, which is separated from $w = 0$ by the image of the arc $|z - z_n| = \eta$ by $w = f_n(z)$. Hence Δ also lies in $|w - z_n| < 3\eta$. Thus we have finally

$$|f_n(z) - z_n| \leqslant 3\eta \quad \text{in} \quad |z - z_n| < \eta,$$

and so $\qquad |f_n(z) - z| < 4\eta \quad \text{for} \quad |z_n - z| \leqslant \eta.$

This, together with (6.8), gives (6.5) since η is arbitrary.

Next (6.7) shows that if r is fixed, $0 < r < 1$, then $\psi_n(z) \to 0$ $(n \to \infty)$, uniformly for $|z| \leqslant r$, and so we have as $n \to \infty$ uniformly for $|z| \leqslant r$,

$$f_n(z) - z = z[e^{\psi_n(z)} - 1] \sim z\psi_n(z)$$

$$\sim z \log \beta_n \frac{z_n + z}{z_n - z}$$

by (6.7), and since $\beta_n \to 1$ $(n \to \infty)$, we deduce (6.6). This proves Lemma 6.5.

6.4. Löwner's slit mappings. Let γ be a simple Jordan arc having one end-point B on $|w| = 1$ and lying otherwise in $|w| < 1$. We suppose further than γ does not pass through $w = 0$ and consists of a finite number of analytic arcs $P_1 P_2$, $P_2 P_3$, ..., $P_{n-1} B$. We shall call such an arc γ a sectionally analytic cut, s.a. cut for short.

The set of points G consisting of all points of $|w| < 1$ not on γ will be a simply connected domain. In fact if Q_1, Q_2 lie in G near different points of γ, we can pass from Q_1 to Q_2 in G by a curve near γ, which if necessary will go round the tip of γ and along the other side. Thus any two points of G near γ can be joined in G, and any point in G can be joined to some point near γ, for instance, by a straight-line segment to the tip of γ. Thus G is connected. Further, the complement of G consists of $|w| \geqslant 1$, together with γ and so is closed and connected. Thus† G is a simply connected bounded domain containing $w = 0$, and so by Riemann's mapping theorem‡ there exists a unique function or transformation

$$w = f(z) = \beta(z + a_2 z^2 + \dots), \tag{6.9}$$

mapping $|z| < 1$ $(1, 1)$ conformally onto G so that $f(0) = 0$ and $f'(0) = \beta > 0$. Clearly $f(z)/\beta \in \mathfrak{S}$. We have further

LEMMA 6.7. *The functions $f(z)/\beta$, where $f(z)$ is constructed as in (6.9), form a dense subclass $\mathfrak{S}''$ of $\mathfrak{S}$.*

Suppose that $f(z) \in \mathfrak{S}$ and $0 < \rho < 1$. Then $f(\rho z)$ maps $|z| < 1$ onto the interior of an analytic curve, namely, the image of

---

† Ahlfors[2], p. 112.   ‡ Ahlfors[2], p. 172.

$|z| = \rho$ by $f(z)$. Also $\dfrac{1}{\rho} f(\rho z) \in \mathfrak{S}$, $\dfrac{1}{\rho} f(\rho z) \to f(z)$ as $\rho \to 1$, uniformly for $|z| \leqslant r$, when $0 < r < 1$.

It is thus sufficient to show that the functions $\dfrac{1}{\rho} f(\rho z)$ can be approximated by functions in $\mathfrak{S}''$. Next if M is large

$$w = \psi(z) = \frac{1}{\rho M} f(\rho z) = \frac{1}{M} z + \ldots$$

maps $|z| < 1$ onto the interior D of a closed analytic curve Γ lying entirely in $|w| < 1$. Let Γ_n consist of a straight line segment

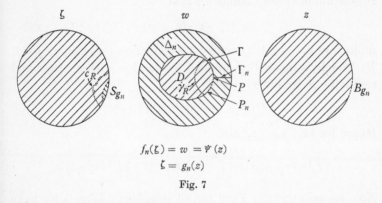

$$f_n(\zeta) = w = \Psi(z)$$
$$\zeta = g_n(z)$$

Fig. 7

from $w = 1$ to the nearest point P of Γ and then along Γ until the whole of Γ except an arc $P_n P$ of diameter $1/n$ is described. Let Δ_n consist of $|w| < 1$ except for the s.a. cut Γ_n and let

$$f_n(z) = \beta_n(z + a_2 z^2 + \ldots)$$

map $|z| < 1$ onto Δ_n. It remains to show that $f_n(z) \to \psi(z)$ $(n \to \infty)$, and so $f_n(z)/\beta_n$ approximates $\dfrac{1}{\rho} f(\rho z)$ for a fixed ρ, which in turn approximates $f(z)$.

Consider now $\zeta = f_n^{-1}[\psi(z)] = g_n(z)$, say. We verify that $g_n(0) = 0$, $g_n'(0) > 0$, so that $g_n(z)$ is a transformation. Let S_{g_n} be defined as in § 6.2. Then S_{g_n} consists of all those points in $|\zeta| < 1$ which correspond to points outside D by the transformation $w = f_n(\zeta)$.

Choose now δ so small that the circle of centre P and radius R meets D in a single arc γ_R for $0 < R < \delta$ and that the origin $w = 0$ lies outside this circle. Then if $1/n < R$, γ_R corresponds to a single arc c_R in $|\zeta| < 1$ by $\zeta = f_n^{-1}(w)$ and all points of S_{g_n} are separated from $\zeta = 0$ by c_R. By Lemma 6.1 we may choose R so that the length $l(R)$ of c_R satisfies

$$l(R) \leqslant \pi[\tfrac{1}{2}\log(n\delta)]^{-\frac{1}{2}},$$

and so by Lemma 6.2 the diameter $d(g_n)$ of S_{g_n} satisfies

$$d(g_n) \leqslant \pi[\tfrac{1}{2}\log(n\delta)]^{-\frac{1}{2}}.$$

It now follows from Lemma 6.6 that

$$g_n(z) \to z \quad (n \to \infty),$$

uniformly in $|z| < 1$. Given $r < 1$, choose now ρ so that $r < \rho < 1$. By Cauchy's inequality, applied to the disc with centre z and radius $1 - \rho$, we have

$$|f_n'(z)| \leqslant (1-\rho)^{-1} \quad (|z| \leqslant \rho).$$

Hence for $|z_1| \leqslant \rho$, $|z_2| \leqslant \rho$,

$$|f_n(z_1) - f_n(z_2)| = \left| \int_{z_1}^{z_2} f_n'(z)\,dz \right| \leqslant (1-\rho)^{-1} |z_1 - z_2|.$$

Now suppose $|z| \leqslant r$. Then for all sufficiently large n we have $|g_n(z)| \leqslant \rho$, since $g_n(z) \to z$ uniformly. Hence

$$|f_n(z) - \psi(z)| = |f_n(z) - f_n[g_n(z)]|$$
$$\leqslant (1-\rho)^{-1} |z - g_n(z)|,$$

and so $f_n(z) \to \psi(z)$ $(n \to \infty)$ uniformly for $|z| \leqslant r$. This completes the proof of Lemma 6.7.

6.5. Continuity properties.

Following Löwner we now investigate the class $\mathfrak{S}''$ and show that the functions $f(z)$ in (6.9) can be obtained by a series of successive infinitesimal transformations from $w = z$.

Let γ be an s.a. cut given by $w = \alpha(t)$ $(0 \leqslant t \leqslant t_0)$, where $\alpha(t) \neq 0$, $|\alpha(t)| < 1$ $(0 \leqslant t < t_0)$ and $|\alpha(t_0)| = 1$. We denote by $\gamma_{t't''}$ the arc $t' \leqslant t \leqslant t''$ of γ and by γ_t the arc γ_{tt_0}. Let $G(t)$ consist of $|w| < 1$, except for γ_t. As t increases from 0 to t_0, $G(t)$ expands from $G = G(0)$ to $|w| < 1$. We denote by

$$w = g_t(z) = \beta(t)\,(z + a_2(t)\,z^2 + \ldots) \quad (\beta(t) > 0),$$

the function which maps $|z| < 1$ onto $G(t)$ and aim to show that $g_t(z)$ varies continuously with t as t increases from $g_0(z) = f(z)$ to $g_{t_0}(z) = z$. We have first

LEMMA 6.8. *If $w = g_t(z)$ is defined as above, then the inverse function $z = g_t^{-1}(w)$ remains continuous at $w = \alpha(t)$. Thus as $w = g_t(z) \to \alpha(t)$ in any manner from $G(t)$, z approaches a point $\lambda(t)$ such that $|\lambda(t)| = 1$.*

Choose δ so small that the circle $|w - \alpha(t)| = R$ meets γ_t in exactly one point for $0 < R < \delta$. The choice is possible since γ_t has a continuous tangent at $\alpha(t)$. Then for $0 < R < \delta$ the circle $|w - \alpha(t)| = R$ lies in $G(t)$ except for a single point. The image of this circle is an arc c_R lying, except for end-points, in $|z| < 1$. By Lemma 6.1 we can then choose a sequence $R_n \to 0$ such that $l(R_n) \to 0$, where $l(R_n)$ is the length of c_{R_n}. If $R_n < |\alpha(t)|$, then the disc $|w - \alpha(t)| < R_n$ cut along the arc γ_t corresponds to one of the domains into which c_{R_n} divides $|z| < 1$, namely, that one, Δ_n, which does not contain $z = 0$. By Lemma 6.2 the diameter of Δ_n is not greater than $l(R_n)$ and so tends to zero as $n \to \infty$. Since $\Delta_n \subset \Delta_{n-1}$ when $R_n < R_{n-1}$, it follows that Δ_n shrinks to a single point $\lambda(t)$ as $n \to \infty$.

Thus we may choose n so large that Δ_n lies in the disc $|z - \lambda(t)| < \epsilon$. Then if $|w - \alpha(t)| < R_n$ and $w = g_t(z)$, we have $|z - \lambda(t)| < \epsilon$ and this proves Lemma 6.8.

6.5.1. We next write

$$h(z, t', t'') = g_{t''}^{-1}[g_{t'}(z)] \quad (0 \leqslant t' < t'' \leqslant t_0).$$

We shall study the behaviour of $h(z, t', t'')$, as $t'' - t' \to 0$.

We note that $w = g_{t'}(z)$ maps $|z| < 1$ onto $G(t')$, that is, $|w| < 1$ cut along $\gamma_{t'}$. Also $\zeta = g_{t''}^{-1}(w)$ maps $G(t'')$ onto $|\zeta| < 1$, and so $G(t')$ corresponds to $|\zeta| < 1$ except for the image of $\gamma_{t't''}$ by $\zeta = g_{t''}^{-1}(w)$. For $\gamma_{t't''}$ lies, except for one end-point $\alpha(t'')$, in $G(t'')$

but not in $G(t')$. By Lemma 6.8, $\zeta = g_{t''}^{-1}(w)$ is continuous also at $\alpha(t'')$ and so on the whole of $\gamma_{t't''}$ and maps $\gamma_{t't''}$ onto a Jordan arc $S_{t't''}$ in $|\zeta| \leqslant 1$.

Thus $\zeta = h(z, t', t'')$ maps $|z| < 1$ onto $|\zeta| < 1$ cut along $S_{t't''}$. We write $B_{t't''}$ for the set of boundary points on $|z| = 1$ which correspond to $S_{t't''}$ by this transformation. Other points of

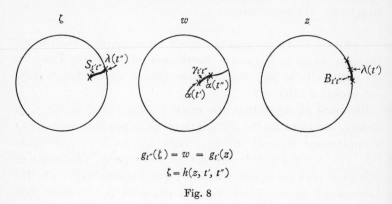

$$g_{t''}(\zeta) = w = g_{t'}(z)$$
$$\zeta = h(z,\, t',\, t'')$$

Fig. 8

$|z| = 1$ correspond only to $|\zeta| = 1$. We note also that $z = \lambda(t')$ belongs to $B_{t't''}$ and corresponds to the tip of the cut $S_{t't''}$ and that the other end-point of $S_{t't''}$ on $|\zeta| = 1$ is $\zeta = \lambda(t'')$.

We now have

LEMMA 6.9. *As $t'' - t' \to 0$, while either $t = t'$ or $t = t''$ remains fixed, both $S_{t't''}$ and $B_{t't''}$ approach the point $\lambda(t)$ in their respective planes. Also $\lambda(t)$ is continuous.*

Suppose first that $t'' \to t'$ while t' remains fixed. Then the arc $\gamma_{t't''}$ shrinks to the fixed point $\alpha(t')$ and so finally lies in a disc of centre $\alpha(t')$ and radius δ. If ϵ is sufficiently small it follows from Lemma 6.8 that we can choose δ depending on ϵ, such that if $w = g_{t'}(z)$ and $|w - \alpha(t')| < \delta$, then $|z - \lambda(t')| < \epsilon$. Thus if $\gamma_{t't''}$ has diameter less than δ, $B_{t't''}$ has diameter less than ϵ. Thus as $t'' \to t'$, the diameter of $B_{t't''}$ tends to zero and hence so does that of $S_{t't''}$ by Lemma 6.4.

Similarly, if $t' \to t''$, while t'' remains fixed, $\gamma_{t't''}$ shrinks to the point $\alpha(t'')$. It then follows from the continuity of $g_{t''}^{-1}(w)$ at $w = \alpha(t'')$, that $S_{t't''}$, which corresponds to $\gamma_{t't''}$ by $\zeta = g_{t''}^{-1}(w)$, shrinks to the point $g_{t''}^{-1}[\alpha(t'')] = \lambda(t'')$, and so the diameter of $S_{t't''}$ tends to zero. Thus the diameter of $B_{t't''}$ tends to zero also by Lemma 6.4.

It now follows from Lemma 6.6 that in either case

$$h(z, t', t'') \to z \tag{6.10}$$

uniformly in $|z| < 1$ as $t'' - t' \to 0$ through any sequence of values. Thus if z lies on $B_{t't''}$ and ζ is a corresponding point on $S_{t't''}$, we have $|z - \zeta| < \epsilon$ finally.

Also $S_{t't''}$ always contains $\lambda(t'')$ and $B_{t't''}$ contains $\lambda(t')$. Taking $z = \lambda(t')$ choose t'' so near t' that the diameter of $S_{t't''}$ is less than ϵ. Then $|\zeta - \lambda(t'')| < \epsilon$ and $|\zeta - z| < \epsilon$, and so

$$|\lambda(t'') - \lambda(t')| < 2\epsilon.$$

Similarly, if t'' is fixed choose $\zeta = \lambda(t'')$ and t' so near t'' that the diameter of $B_{t't''}$ is less than ϵ. Then $|\zeta - z| < \epsilon$ and $|z - \lambda(t')| < \epsilon$, and so $|\lambda(t'') - \lambda(t')| < 2\epsilon$. Thus in either case $\lambda(t'') - \lambda(t') \to 0$, and if $t = t'$ or t'' and t is fixed $\lambda(t')$ and $\lambda(t'')$ approach $\lambda(t)$. Thus $\lambda(t)$ is continuous. Since the diameters of $S_{t't''}$, $B_{t't''}$ tend to zero and these sets contain $\lambda(t'')$ and $\lambda(t')$ respectively, both these sets approach $\lambda(t)$ as $t'' - t' \to 0$, while one of t', t'' remains fixed. This proves Lemma 6.9.

6.6. The differential equation. We note that for $0 \leqslant t' < t'' \leqslant t_0$,

$$h(z, t', t'') = \frac{\beta(t')}{\beta(t'')} z + \dots$$

satisfies the conditions of Schwarz's lemma in $|z| < 1$. Thus $\beta(t)$ is a strictly increasing function of t in $0 \leqslant t \leqslant t_0$. Again by (6.10)

$$\frac{\beta(t')}{\beta(t'')} \to 1 \quad (t'' - t' \to 0),$$

while either t' or t'' remains fixed. Thus $\beta(t)$ is continuous. Hence

$$\tau = \log \frac{\beta(t)}{\beta(0)}$$

is a continuous strictly increasing function of t for $0 \leqslant t \leqslant t_0$, and we may take τ for our parameter t, which has been left undetermined so far. We shall do so in what follows. With this normalization

$$g_t(z) = \beta e^t (z + a_2(t) z^2 + \dots) \quad (0 \leqslant t \leqslant t_0), \tag{6.11}$$

9

where $\beta = \beta(0) = f'(0)$, $t_0 = \log(1/\beta)$. Also

$$h(z, t', t'') = e^{t'-t''} z + \dots$$

We now prove

LEMMA 6.10. *With the normalization* (6.11) *let* $t'' - t' \to 0$ *while* $t = t'$ *or* t'' *remains fixed in the range* $0 \leqslant t' < t'' \leqslant \log(1/\beta)$. *Then we have, uniformly for* $|z| \leqslant r$, *when* $0 < r < 1$,

$$\frac{h(z, t', t'') - z}{t'' - t'} \to -z \frac{1 + \kappa(t) z}{1 - \kappa(t) z},$$

where $\kappa(t) = \lambda(t)^{-1}$.

In fact Lemmas 6.6 and 6.9 show that, for $|z| \leqslant r < 1$,

$$h(z, t', t'') - z \sim -(1 - e^{t'-t''}) z \frac{e^{i\phi} + z}{e^{i\phi} - z},$$

as $t'' - t' \to 0$, where $e^{i\phi}$ is a point of $B_{t't''}$ so that $e^{i\phi} \to \lambda(t)$. Now Lemma 6.10 follows.

We can now prove part of our fundamental theorem.

THEOREM 6.2. *If* $f(z, t) = g_t^{-1}[f(z)]$ $(0 \leqslant t \leqslant t_0 = \log(1/\beta))$, *then* $f(z, t)$ *satisfies the differential equation*

$$\frac{\partial f}{\partial t} = -f \frac{1 + \kappa(t) f}{1 - \kappa(t) f},$$

with the initial condition $f(z, 0) = z$. *The functions* $\beta^{-1} f(z, t_0)$ *form a dense subclass of* $\mathfrak{S}$.

The function $w = f(z)$ maps $|z| < 1$ onto $|w| < 1$ cut along γ. Also $\zeta = g_t^{-1}(w)$ maps $|w| < 1$ cut along the smaller set γ_t onto $|\zeta| < 1$. Thus $\zeta = f(z, t)$ maps $|z| < 1$ $(1, 1)$ conformally onto a subset of $|\zeta| < 1$ and so is a univalent function for

$$0 \leqslant t \leqslant t_0 = \log(1/\beta).$$

Also near $z = 0$

$$w = f(z) = \beta z + \dots = g_t(\zeta) = \beta e^t \zeta + \dots.$$

Thus $\zeta = e^{-t} z + \dots$ and so $e^t f(z, t) \in \mathfrak{S}$ and in particular

$$\beta^{-1} f(z, t_0) = \beta^{-1} f(z) \in \mathfrak{S}.$$

By Lemma 6.7 this class of functions is dense in $\mathfrak{S}$.

Next write $f(z, t')$ instead of z in Lemma 6.10. This is permissible since $|f(z, t')| < 1$ for $|z| < 1$. Also

$$h[f(z, t'), t', t''] = g_{t''}^{-1}\{g_{t'}(g_{t'}^{-1}[f(z)])\} = g_{t''}^{-1}[f(z)] = f(z, t'').$$

Thus Lemma 6.10 gives

$$\frac{f(z, t'') - f(z, t')}{t'' - t'} \sim -f(z, t')\frac{1 + \kappa(t)f(z, t')}{1 - \kappa(t)f(z, t')} \tag{6.12}$$

if $t'' - t' \to 0$, while t' or t'' remains fixed. If $t = t'$ is fixed this gives the required result for the right derivative. If t'' remains fixed the required result for the left derivative also follows. In fact by Lemma 6.9 and since $\kappa(t) = 1/\lambda(t)$

$$\kappa(t') \to \kappa(t'') \quad (t' \to t''),$$

while t'' is fixed, and by (6.10)

$$h(z, t', t'') - z \to 0 \quad (t' \to t''),$$

uniformly in $|z| < 1$. Writing $f(z, t')$ instead of z we obtain

$$f(z, t'') - f(z, t') \to 0 \quad (t' \to t''),$$

uniformly in $|z| < 1$, so that (6.12) gives, also if $t = t''$ is fixed,

$$\frac{f(z, t) - f(z, t')}{t - t'} \to -f(z, t)\frac{1 + \kappa(t)f(z, t)}{1 - \kappa(t)f(z, t)}$$

as $t' \to t$ from below. This proves Theorem 6.2.

6.7. Completion of proof of Theorem 6.1.

In order to complete the proof of Theorem 6.1 it remains to show that given $\kappa(t)$, continuous and satisfying $|\kappa(t)| = 1$ $(0 \leqslant t \leqslant t_0)$, there exists a unique solution $w = f(z, t)$ of the differential equation (6.1), such that $f(z, 0) = z$, and further that $e^t f(z, t) \in \mathfrak{S}$. The proof of this is a standard piece of work in the theory of differential equations, but we include it for completeness.

It is convenient to put

$$W = U + iV = \log\left(\frac{1}{w}\right).$$

Then (6.1) becomes

$$\frac{\partial W}{\partial t} = \frac{1 + \kappa(t)e^{-W}}{1 - \kappa(t)e^{-W}} = \phi(t, W), \tag{6.13}$$

say. We note that $\Re\phi(t, W) > 0$ for $\Re W > 0$. Also if $U > \delta > 0$,

$$\left|\frac{\partial\phi}{\partial W}\right| = \frac{|\,2e^{-W}\,|}{|\,1 - \kappa(t)\,e^{-W}\,|^2} \leqslant \frac{2}{(1 - e^{-\delta})^2} = K,$$

say, and
$$|\,\phi(t, W)\,| \leqslant \frac{2}{1 - e^{-\delta}} < K. \tag{6.14}$$

If we integrate along the straight line segment from W_1 to W_2 we deduce, if $\Re W_1 > \delta$, $\Re W_2 > \delta$,

$$|\,\phi(t, W_2) - \phi(t, W_1)\,| \leqslant \int_{W_1}^{W_2} \left|\frac{\partial\phi}{\partial W}\right| |\,dW\,| \leqslant K\,|\,W_2 - W_1\,|. \tag{6.15}$$

Thus $\phi(t, W)$ is Lip in W in the half-plane $U > \delta$ uniformly with respect to t.

We now define a sequence of function $F_n(t, \omega)$ $(n \geqslant 0)$ as follows. We suppose $\Re\omega > \delta > 0$ and set

$$\left.\begin{aligned} F_0(t, \omega) &\equiv \omega, \\ F_n(t, \omega) &= \omega + \int_0^t \phi[\tau, F_{n-1}(\tau, \omega)]\,d\tau \quad (0 \leqslant t \leqslant t_0,\ n \geqslant 1). \end{aligned}\right\} \tag{6.16}$$

We note first that $\Re F_n(t, \omega)$ increases with t for fixed ω and so remains greater than δ. For if this is true for n, then

$$\Re\phi[\tau, F_n(\tau, \omega)] > 0$$

and the result remains true for $n + 1$.

Next we have

$$|\,F_n(t, \omega) - F_{n-1}(t, \omega)\,| \leqslant \frac{K^n t^n}{n!} \quad (0 \leqslant t \leqslant t_0,\ n \geqslant 1).$$

In fact if $n = 1$ we have from (6.14)

$$F_1(t, \omega) - F_0(t, \omega) = \left|\int_0^t \phi(\tau, \omega)\,d\tau\right| \leqslant \int_0^t K\,d\tau = Kt.$$

Also if our result is true for n, then (6.15) gives

$$|\,F_{n+1}(t, \omega) - F_n(t, \omega)\,| = \left|\int_0^t \{\phi[\tau, F_n(\tau, \omega)] - \phi[\tau, F_{n-1}(\tau, \omega)]\}\,d\tau\right|$$

$$\leqslant K\int_0^t |\,F_n(\tau, \omega) - F_{n-1}(\tau, \omega)\,|\,d\tau \leqslant K\int_0^t \frac{K^n \tau^n}{n!}\,d\tau = \frac{K^{n+1} t^{n+1}}{(n+1)!}.$$

Thus the result is true for $n + 1$.

It follows that the sequence $F_n(t, \omega)$ converges uniformly for $\Re\omega > \delta$ and $0 \leqslant t \leqslant t_0$ to a limit function $F(t, \omega)$. Taking the limit in (6.16) we deduce

$$F(t, \omega) = \omega + \int_0^t \phi[\tau, F(\tau, \omega)] \, d\tau.$$

Thus $W(t) = F(t, \omega)$ satisfies the differential equation (6.13) with the initial condition $W(0) = \omega$.

We note next that the functions $F_n(t, \omega)$ are analytic in ω for $\Re\omega > 0$ and fixed t and continuous in t, ω jointly. Hence so is their uniform limit $F(t, \omega)$.

Suppose next that $W(t)$ is any solution of (6.13) which satisfies $\Re W(0) > \delta$ and so $\Re W(t) > \delta$ $(0 \leqslant t \leqslant t_0)$ and further that

$$W(t_1) = F(t_1, \omega)$$

for some pair (t_1, ω), such that $0 \leqslant t_1 \leqslant t_0$ and $\Re\omega > \delta$. Then $W(t) \equiv F(t, \omega)$ $(0 \leqslant t \leqslant t_0)$. In fact let

$$M = \sup_{0 \leqslant t \leqslant t_0} |W(t) - F(t, \omega)|.$$

Then (6.13) gives, for $0 \leqslant t \leqslant t_0$,

$$W(t) - W(t_1) = \int_{t_1}^t \phi[\tau, W(\tau)] \, d\tau,$$

$$F(t, \omega) - F(t_1, \omega) = \int_{t_1}^t \phi[\tau, F(\tau, \omega)] \, d\tau,$$

and so

$$|F(t, \omega) - W(t)| = \left| \int_{t_1}^t \{\phi[\tau, W(\tau)] - \phi[\tau, F(t, \omega)]\} \, d\tau \right|. \quad (6.17)$$

We deduce by induction that

$$|F(t, \omega) - W(t)| \leqslant \frac{M K^n (t - t_1)^n}{n!}$$

for every positive integer n. For this holds for $n = 0$ by hypothesis. If it is true for n, then (6.15) and (6.17) give

$$|F(t, \omega) - W(t)| \leqslant K \left| \int_{t_1}^t M K^n (\tau - t_1)^n \, d\tau \right| = \frac{M K^{n+1} |t - t_1|^{n+1}}{(n+1)!}$$

as required. Since n is arbitrary $W(t) \equiv F(t, \omega)$.

Taking $t_1 = 0$, we see that $F(t, \omega)$ is the unique solution of

(6.13) such that $F(0, \omega) = \omega$. We also see that $F(t, \omega)$ is a univalent function of ω for $\Re\omega > 0$ and a fixed positive t. For, if

$$F(t, \omega_1) = F(t, \omega_2),$$

the above uniqueness theorem shows that $F(0, \omega_1) = F(0, \omega_2)$, i.e. $\omega_1 = \omega_2$. Again, if $\omega_2 - \omega_1 = 2\pi i$, we easily see that

$$F(t, \omega_2) - F(t, \omega_1) = 2\pi i.$$

Thus $f(z, t)$, given by

$$-\log f(z, t) = F(t, -\log z) \quad (0 \leqslant t \leqslant t_0),$$

is for fixed t an analytic one-valued univalent function of z for $0 < |z| < 1$, and is the unique solution of the differential equation (6.1), which satisfies $f(z, 0) = z$. Also $|f(z, t)| < 1$ in $|z| < 1$, and so $f(z, t)$ remains regular at $z = 0$ and is clearly zero there. If we apply (6.1) to $g(t) = f(z, t)/z$, we obtain

$$\frac{\partial g(t)}{\partial t} = -g \frac{1 + z\kappa(t)g}{1 - z\kappa(t)g},$$

and putting $z = 0$, we deduce

$$\frac{\partial g}{\partial t} = -g, \quad g(t) = e^{-t}g(0) = e^{-t}.$$

Thus $f(z, t) = ze^{-t} + \dots$ near $z = 0$ and so $e^t f(z, t) \in \mathfrak{S}$. This completes the proof of Theorem 6.1.

6.8. The third coefficient.

We proceed to give a number of applications of Theorem 6.1. Let $\mathfrak{S}$ denote as usual the class of functions
$$f(z) = z + a_2 z^2 + \dots$$

which are univalent in $|z| < 1$. In this section we prove

THEOREM 6.3. *If $f(z) \in \mathfrak{S}$, then $|a_3| \leqslant 3$.*[†]

By Theorem 6.1 we may confine ourselves to the functions $e^{t_0}f(z, t_0)$ of that theorem, since they form a dense subclass of $\mathfrak{S}$. We write $\beta = e^{-t_0}$,

$$f(z, t_0) = \beta\left(z + \sum_{n=2}^{\infty} a_n z^n\right),$$

[†] A rather more detailed discussion shows that $|a_3| < 3$ except when $f(z) = z(1 - z\, e^{i\theta})^{-2}$. A similar remark applies to Theorem 6.4. See Löwner[2].

and proceed to develop Löwner's formulae for a_n in terms of $\kappa(t)$. It is convenient to work with the function $g_t(\zeta)$ defined by

$$g_t[f(z,t)] = f(z,t_0) \quad (0 \leqslant t \leqslant t_0).$$

We write $\zeta = f(z,t)$ and differentiate the above relation with respect to t. This gives

$$\frac{\partial g}{\partial \zeta}\frac{\partial \zeta}{\partial t} + \frac{\partial g}{\partial t} = 0.$$

Substituting in (6.1) we obtain

$$\frac{\partial g}{\partial t} = \frac{\partial g}{\partial \zeta}\zeta\frac{1+\kappa\zeta}{1-\kappa\zeta}. \tag{6.18}$$

We have $f(z,t) = e^{-t}(z + \dots)$ near $z = 0$ and so

$$g_t(\zeta) = \beta\, e^t\left[\zeta + \sum_{n=2}^{\infty} c_n(t)\,\zeta^n\right]$$

near $\zeta = 0$. We substitute in (6.18) and obtain for $n \geqslant 2$

$$c_n'(t) + c_n(t) = nc_n(t) + 2\sum_{r=1}^{n-1}(n-r)\,c_{n-r}(t)\,[\kappa(t)]^r,$$

i.e. $\quad c_n'(t) = (n-1)\,c_n(t) + 2\sum_{r=1}^{n-1} rc_r(t)\,[\kappa(t)]^{n-r} \quad (0 \leqslant t \leqslant t_0).$

The term-by-term differentiation with respect to t may be justified by expressing the coefficients as contour integrals in terms of $g_t(\zeta)$. Also when $t = 0$, $f(z,0) \equiv z$, $g_0(z) \equiv f(z,t_0)$, $c_n(0) \equiv a_n$. When $t = t_0$, $g_t(z) \equiv z$ and so $c_n(t_0) = 0$ $(n \geqslant 2)$. Finally, $c_1(t) \equiv 1$. Thus we have the boundary conditions

$$c_1(t) \equiv 1 \quad c_n(0) = a_n, \quad c_n(t_0) = 0 \quad (n \geqslant 2).$$

From this system we can successively determine the coefficients. We obtain in fact

$$c_n(t) = -2e^{(n-1)t}\int_t^{t_0}\left\{\sum_{r=1}^{n-1} rc_r(\tau)\,[\kappa(\tau)]^{n-r}\right\}e^{-(n-1)\tau}\,d\tau \quad (0 \leqslant t \leqslant t_0).$$

This gives $\qquad c_2(t) = -2e^t\int_t^{t_0}\kappa(\tau)\,e^{-\tau}\,d\tau.$

Also

$$c_3(t) = -2e^{2t}\left\{ \int_t^{t_0} \kappa(\tau)^2 e^{-2\tau}\,d\tau + 2\int_t^{t_0} \kappa(\tau)\,c_2(\tau)\,e^{-2\tau}\,d\tau \right\}$$

$$= -2e^{2t}\left\{ \int_t^{t_0} [\kappa(\tau)]^2 e^{-2\tau}\,d\tau - 2\left(\int_t^{t_0} \kappa(\tau)\,e^{-\tau}\,d\tau \right)^2 \right\},$$

since

$$\frac{d}{dt}\left(\int_t^{t_0} \kappa(\tau)\,e^{-\tau}\,d\tau \right)^2 = -2\kappa(t)\,e^{-t}\int_t^{t_0} \kappa(\tau)\,e^{-\tau}\,d\tau = \kappa(t)\,c_2(t)\,e^{-2t}.$$

We deduce at once that $|a_2| = |c_2(0)| \leqslant 2$. The find the upper bound for $|a_3| = |c_3(0)|$, we may without loss in generality suppose a_3 real and positive, since this may be achieved by considering $e^{-i\phi}f(z\,e^{i\phi})$ instead of $f(z)$. We write $\kappa(t) = e^{i\theta(t)}$ and consider

$$\Re a_3 = -2\int_0^{t_0} [2\cos^2\theta(t) - 1]\,e^{-2t}\,dt + 4\left(\int_0^{t_0} \cos\theta(t)\,e^{-t}\,dt \right)^2$$

$$-4\left(\int_0^{t_0} \sin\theta(t)\,e^{-t}\,dt \right)^2.$$

To obtain an upper bound for $\Re a_3$ we omit the third term on the right. The first term is

$$-4\int_0^{t_0} e^{-2t}\cos^2\theta(t)\,dt + 1 - e^{-2t_0},$$

and by Schwarz's inequality the second term is at most

$$4\int_0^{t_0} e^{-t}\,dt \int_0^{t_0} e^{-t}\cos^2\theta(t)\,dt < 4\int_0^{t_0} e^{-t}\cos^2\theta(t)\,dt.$$

Thus we obtain

$$|a_3| = \Re a_3 \leqslant 1 + 4\int_0^{t_0} \cos^2\theta(t)\,(e^{-t} - e^{-2t})\,dt$$

$$\leqslant 1 + 4\int_0^{\infty} (e^{-t} - e^{-2t})\,dt = 3.$$

This proves Theorem 6.3.

6.9. Coefficients of the inverse functions. Suppose again that $w = f(z) \in \mathfrak{S}$ and let the inverse function

$$z = \phi(w) = f^{-1}(w)$$

be given by

$$\phi(w) = w + \sum_{m=2}^{\infty} b_m w^m$$

near $w = 0$. By means of Theorem 6.1 Löwner obtained the exact bounds of all the coefficients b_m. His result is

THEOREM 6.4. *If* $w = f(z) \in \mathfrak{S}$ *and* $z = f^{-1}(w) = w + \sum\limits_{m=2}^{\infty} b_m w^m$ *then*

$$|b_m| \leqslant \frac{1.3.5.\ldots.(2m-1)\,2^m}{(m+1)!} \quad (m \geqslant 2).$$

Equality holds when

$$f(z) = \frac{z}{(1+z)^2}, \quad f^{-1}(w) = [1 - 2w - (1 - 4w)^{\frac{1}{2}}]/(2w).$$

It is again sufficient to consider instead of $f(z)$ the functions $e^{t_0} f(z, t_0)$ of Theorem 6.1. For these functions can be used to approximate a general $f(z) \in \mathfrak{S}$, so that their coefficients approximate those of $f(z)$, and the coefficients of the inverse functions $f^{-1}(w)$, which are polynomials in the coefficients of $f(z)$, can be similarly approximated.

Let then $f(z, t)$ be the function of Theorem 6.1 and let $z = \phi_t(w)$ be the inverse function so that

$$\phi_t[f(z, t)] = z. \tag{6.19}$$

We write $\quad \phi_t(w) = e^t \left[w + \sum\limits_{m=2}^{\infty} b_m(t)\, w^m \right] \quad (0 \leqslant t \leqslant t_0), \tag{6.20}$

and $\beta = e^{-t_0}$. Thus

$$\phi(w) = \phi_{t_0}(\beta w) = w + \sum\limits_{m=2}^{\infty} b_m w^m$$

is inverse to $\beta^{-1} f(z, t) \in \mathfrak{S}$, and so we need only prove our inequalities for the coefficients b_m of this function $\phi(w)$, where

$$b_m = \beta^{m-1} b_m(t_0).$$

We note that the equations (6.1) and (6.19) lead again to the analogue of (6.18), namely,

$$\frac{\partial \phi_t(w)}{\partial t} = \frac{\partial \phi_t(w)}{\partial w} w \frac{1 + \kappa(t)\, w}{1 - \kappa(t)\, w}.$$

Substituting from (6.20) we obtain just as in the previous section

$$b'_m(t) + b_m(t) = m b_m(t) + 2 \sum\limits_{r=1}^{m-1} r b_r(t) \,[\kappa(t)]^{m-r} \quad (0 \leqslant t \leqslant t_0, \ m \geqslant 2),$$

with the boundary conditions

$$b_1(t) \equiv 1 \quad (0 \leqslant t \leqslant t_0), \qquad b_m(0) = 0, \quad b_m(t_0) = \beta^{-m+1} b_m \quad (m \geqslant 2).$$

These yield the inductive relation

$$b_m(t) = 2e^{(m-1)t} \int_0^t \left\{ \sum_{r=1}^{m-1} rb_r(\tau) [\kappa(\tau)]^{m-r} \right\} e^{-(m-1)\tau} d\tau \quad (m > 1).$$

It is now clear that $|b_m(t_0)|$ attains its maximum possible value for any fixed $t_0 > 0$ and $m > 1$ if $\kappa(t) \equiv 1$. In this case all the $b_m(t)$ are real and positive. It is also evident that the corresponding value of $b_m = e^{-(m-1)t_0} b_m(t_0)$ increases with increasing t_0 and so we obtain the upper bound for variable t_0 in the limit as $t_0 \to \infty$.

We now take $\kappa(t) \equiv 1$ in the differential equation (6.1) and obtain on integration

$$\frac{f(z, t_0)}{1 + [f(z, t_0)]^2} = \frac{\beta z}{(1+z)^2},$$

i.e.

$$\frac{w}{(1+w)^2} = \frac{\beta \phi_{t_0}(w)}{[1 + \phi_{t_0}(w)]^2}.$$

Writing $\phi(w) = \phi_{t_0}(\beta w)$ we deduce

$$\frac{\phi(w)}{[1 + \phi(w)]^2} = \frac{w}{(1 + \beta w)^2}.$$

Thus $\beta \to 0$ $(t_0 \to \infty)$ and $\phi(w) \to \psi(w)$, where

$$\frac{\psi(w)}{[1 + \psi(w)]^2} = w, \quad \text{i.e.} \quad \psi(w) = \frac{1 - 2w - \sqrt{(1 - 4w)}}{2w} = \sum_{m=1}^{\infty} b_m w^m,$$

and

$$b_m = \frac{1 \cdot 3 \cdots (2m-1) \, 2^m}{(m+1)!},$$

as required. The inverse function $w = \psi^{-1}(z) = z(1+z)^{-2} \in \mathfrak{S}$, and so $\psi(w)$ has the largest coefficients in our class and Theorem 6.4 is proved.

6.10. The argument† of $f(z)/z$. While the elementary methods of Chapter 1 are adequate to obtain the bounds for $|f(z)|$, $|f'(z)|$, etc., when $f(z) \in \mathfrak{S}$, the bounds for $\arg(f(z)/z)$,

---

† If $\phi(z) \neq 0$ in $|z| < 1$ and $\phi(0) > 0$ we define $\arg \phi(z)$ as the imaginary part of $\log \phi(0) + \int_0^z \frac{\phi'(\zeta)}{\phi(\zeta)} \, d\zeta$, where the integral is taken along a straight line segment.

$\arg f'(z)$, etc., lie deeper. Here the function $z(1-z)^{-2}$ is no longer extremal. Following Grunsky [1], we prove

THEOREM 6.5. *Suppose that $f(z) \in \mathfrak{S}$. Then we have*

$$-\log \frac{1+|z|}{1-|z|} \leqslant \arg \frac{f(z)}{z} \leqslant \log \frac{1+|z|}{1-|z|}. \qquad (6.21)$$

Both these inequalities are sharp for any fixed z in $|z| < 1$.

It is again sufficient to consider the functions $e^{t_0} f(z, t_0)$ of Theorem 6.1, and hence $f(z, t_0)$, since $\arg e^{t_0} = 0$. We write $f = f(z, t)$, $\kappa = \kappa(t)$ for short. Then (6.1) gives

$$\frac{\partial}{\partial t} \log f = -\frac{1+\kappa f}{1-\kappa f} = \frac{-(1+\kappa f)(1-\bar{\kappa}\bar{f})}{|1-\kappa f|^2}.$$

Taking real and imaginary parts we deduce

$$\frac{\partial}{\partial t} \log |f| = -\frac{1-|f|^2}{|1-\kappa f|^2}, \qquad (6.22)$$

$$\frac{\partial}{\partial t} \arg f = -\frac{2\Im(\kappa f)}{|1-\kappa f|^2}. \qquad (6.23)$$

The equation (6.22) shows that $|f|$ decreases strictly with increasing t. Also (6.22) and (6.23) together give

$$d_t \arg f \leqslant \frac{2|f|}{1-|f|^2} d_t \log \frac{1}{|f|}. \qquad (6.24)$$

We integrate this from $t = 0$ to t_0 and note that $f(z, 0) = z$. Thus

$$\arg \frac{f(z, t_0)}{z} \leqslant \log \frac{(1+|z|)(1-|f(z, t_0)|)}{(1-|z|)(1+|f(z, t_0)|)} \leqslant \log \frac{1+|z|}{1-|z|}.$$

This yields the upper bound in (6.21). The lower bound follows similarly. It is worth noting that the methods of Chapter 1 only lead to

$$\left| \arg \frac{f(z)}{z} \right| \leqslant 2 \log \frac{1+|z|}{1-|z|}.$$

We next show that the right-hand inequality of (6.21) is sharp for a fixed value z_0 of z in $|z| < 1$. To do this we have to find $\kappa(t)$ in an assigned range $0 \leqslant t \leqslant t_0$ so that the solution $f = f(z_0, t)$ of the differential equation

$$\frac{df}{dt} = -f \frac{1 + \kappa f}{1 - \kappa f} \quad (0 \leqslant t \leqslant t_0),$$

with the initial condition $f(z_0, 0) = z_0$ satisfies

$$\Im(\kappa f) = -|f|.$$

For in this case we shall have equality in (6.24). Also (6.22) gives in this case

$$\frac{\partial}{\partial t} \log |f| = -\frac{1 - |f|^2}{1 + |f|^2},$$

so that $f = f(z_0, t)$ satisfies

$$\frac{|f|}{1 - |f|^2} = e^{-t} \frac{|z_0|}{1 - |z_0|^2}.$$

We define $|f| = |f(z_0, t)|$ by this equation, then $\arg f(z_0, t)$ by

$$d_t \arg f = -\frac{2 d_t |f|}{1 - |f|^2},$$

so that

$$\arg \frac{f(z_0, t)}{z_0} = \log \left(\frac{1 + |z_0|}{1 - |z_0|} \right) - \log \frac{1 + |f|}{1 - |f|},$$

and finally $\kappa(t)$ by

$$\kappa(t) = \frac{-i |f(z_0, t)|}{f(z_0, t)}.$$

With these definitions of $\kappa(t)$, $f(z_0, t)$ (6.22) and (6.23) are satisfied and $\arg[f(z_0, t)/z_0]$ can be chosen as close as we please to $\log[(1 + |z_0|)/(1 - |z_0|)]$, since $f(z_0, t) \to 0$ $(t \to \infty)$. Thus the solution $e^{t_0} f(z_0, t_0)$ of the equation (6.1) corresponding to this value of $\kappa(t)$ belongs to $\mathfrak{S}$ and approaches the upper bound in (6.21) arbitrarily closely, so that this bound is sharp. By considering $\overline{f(z_0, t)}$ instead of $f(z_0, t)$ we can show that the lower bound is sharp. This completes the proof of Theorem 6.5.

6.11. Radii of convexity and starshapedness.

The function $f(z)$ maps $|z| = r$ onto a convex curve $\gamma(r)$, if the tangent to $\gamma(r)$ at the point $f(r e^{i\theta})$ turns continuously in an anticlockwise direction as θ increases. The condition for this is that $\arg[ir\,e^{i\theta}f'(r\,e^{i\theta})]$ increases with increasing θ, so that

$$\frac{\partial}{\partial\theta}\arg f'(r\,e^{i\theta})+1\geqslant 0 \quad (0\leqslant\theta\leqslant 2\pi),$$

i.e. $\quad \Im\left\{\dfrac{\partial}{\partial\theta}\log f'(r\,e^{i\theta})\right\}=\Re\left\{r\,e^{i\theta}\dfrac{f''(r\,e^{i\theta})}{f'(r\,e^{i\theta})}\right\}\geqslant -1 \quad (0\leqslant\theta\leqslant 2\pi).$

This gives $\qquad\qquad \Re\left\{z\dfrac{f''(z)}{f'(z)}\right\}\geqslant -1 \quad (|z|=r).$

The inequality (1.6) of Chapter 1 gives for $f(z)\in\mathfrak{S}$

$$\Re\left\{z\frac{f''(z)}{f'(z)}\right\}\geqslant\frac{r(2r-4)}{1-r^2} \quad (|z|=r),$$

so that our condition is satisfied if $2r^2-4r\geqslant r^2-1$, i.e.

$$r^2-4r+1\geqslant 0, \quad 0\leqslant r\leqslant 2-\sqrt{3}.$$

Thus $f(z)\in S$ maps $|z|=r$ onto a convex curve for $0\leqslant r\leqslant 2-\sqrt{3}$. On the other hand if $f(z)=z(1-z)^{-2}$, then

$$z\frac{f''(z)}{f'(z)}=\frac{2z^2+4z}{1-z^2},$$

and this is real and less than -1 for $-1<r<\sqrt{3}-2$. Thus this function does not map $|z|=r$ onto a convex curve for $r>2-\sqrt{3}$. The quantity $r_c=2-\sqrt{3}$ is called the *radius of convexity*.[†]

We may ask similarly for the radius of the largest circle $|z|=r$ such that the image $\gamma(r)$ of $|z|=r$ by $f(z)$ always bounds a starshaped domain with respect to $w=0$. The condition for this was seen in Chapter 1, (1.14) to be that

$$\Re\left\{z\frac{f'(z)}{f(z)}\right\}\geqslant 0, \quad \text{i.e.} \quad \left|\arg z\frac{f'(z)}{f(z)}\right|\leqslant\frac{\pi}{2} \quad (|z|=r).$$

If we write $\qquad\qquad \phi(z)=\dfrac{f\left(\dfrac{z_0+z}{1+\bar{z}_0 z}\right)-f(z_0)}{(1-|z_0|^2)f'(z_0)},$

† Gronwall[1].

then $\phi(z) \in \mathfrak{S}$ if and only if $f(z) \in \mathfrak{S}$. On applying the inequality (6.21) to $\phi(z)$ at $z = -z_0$, we obtain

$$\left| \arg \left(\frac{z_0 f(z_0)}{f'(z_0)} \right) \right| \leqslant \log \frac{1 + |z_0|}{1 - |z_0|},$$

and this is sharp since (6.21) is sharp.

Thus the *radius of starshapedness*, r_s, being the radius of the largest circle whose interior is always mapped onto a starshaped domain with respect to $w = 0$ by $f(z) \in \mathfrak{S}$, is given by†

$$\frac{\pi}{2} = \log \frac{1 + r_s}{1 - r_s}, \quad r_s = \tanh \frac{\pi}{4} = 0.65 \dots.$$

6.12. The argument of $f'(z)$. As a final application we prove the following result of Golusin[1].

THEOREM 6.6. *Suppose that* $f(z) \in \mathfrak{S}$. *Then we have the sharp inequalities*

$$|\arg f'(z)| \leqslant 4 \sin^{-1} |z| \quad \left(|z| \leqslant \frac{1}{\sqrt{2}} \right),$$

$$|\arg f'(z)| \leqslant \pi + \log \frac{|z|^2}{1 - |z|^2} \quad \left(\frac{1}{\sqrt{2}} < |z| < 1 \right).$$

We may again confine ourselves to the functions $f(z, t)$ of Theorem 6.1. We start with the equation (6.1)

$$\frac{\partial}{\partial t} f(z, t) = -f \frac{1 + \kappa f}{1 - \kappa f} = f + \frac{2}{\kappa} - \frac{2}{\kappa(1 - \kappa f)},$$

and differentiate both sides with respect to z. This leads to

$$\frac{\partial}{\partial t} f'(z, t) = f'(z, t) \left[1 - \frac{2}{(1 - \kappa f)^2} \right],$$

i.e.

$$\frac{\partial}{\partial t} \log f'(z, t) = 1 - \frac{2}{(1 - \kappa f)^2}.$$

Taking imaginary parts we deduce

$$\frac{\partial}{\partial t} \arg f'(z, t) = \frac{2 \Im(1 - \kappa f)^2}{|1 - \kappa f|^4}.$$

† Grunsky[2].

We eliminate t between this and (6.22) and deduce

$$d_t \arg f' = \frac{2\Im(1-\kappa f)^2}{|1-\kappa f|^2} \cdot \frac{-d_t|f|}{|f|(1-|f|^2)},$$

as t increases.

Now since $|\kappa(t)| = 1$

$$\frac{|\Im(1-\kappa f)^2|}{|1-\kappa f|^2} = |\sin[2\arg(1-\kappa f)]|$$

$$\leqslant \begin{cases} 2|f|\sqrt{(1-|f|^2)}, & \left(|f| \leqslant \dfrac{1}{\sqrt{2}}\right), \\ 1, & \left(|f| \geqslant \dfrac{1}{\sqrt{2}}\right). \end{cases} \qquad (6.25)$$

Recalling that $d_t|f| < 0$, we deduce that

$$d_t \arg f' \leqslant \begin{cases} \dfrac{-4d_t|f|}{\sqrt{(1-|f|^2)}} & \left(|f| \leqslant \dfrac{1}{\sqrt{2}}\right), \\ \dfrac{-2d_t|f|}{|f|(1-|f|^2)} & \left(|f| > \dfrac{1}{\sqrt{2}}\right). \end{cases}$$

Integrating from $t = 0$ to t_0, we obtain for $|z| \leqslant \dfrac{1}{\sqrt{2}}$

$$|\arg f'| \leqslant \int_{|f|}^{|z|} \frac{4dx}{\sqrt{(1-x^2)}} \leqslant 4\sin^{-1}|z|,$$

and for $|z| > \dfrac{1}{\sqrt{2}}$

$$|\arg f'| \leqslant \int_{|f|}^{\frac{1}{\sqrt{2}}} \frac{4dx}{\sqrt{(1-x^2)}} + \int_{\frac{1}{\sqrt{2}}}^{|z|} \frac{2dx}{x(1-x^2)} \leqslant \pi + \log\left(\frac{|z|^2}{1-|z|^2}\right).$$

This proves the inequalities of Theorem 6.6. To show that they are precise we must find $\kappa(t)$, such that if $f(z_0, t)$ is the solution of (6.1) with assigned initial value $f(z_0, 0) = z_0$, then equality holds in the inequalities (6.25). The resulting equation enables us to calculate $\kappa f(z_0, t)$ in terms of $|f(z_0, t)|$, hence $|f(z_0, t)|$ in terms of t by means of (6.22) and then $\arg f(z_0, t)$ in terms of t by means of (6.23). We can then choose $\kappa(t)$, so that equality holds in (6.25). When this is done for $0 \leqslant t \leqslant t_0$, then (6.22) and (6.23) and equality in (6.25) will hold simultaneously and we shall have

$$\arg f'(z_0, t_0) = \int_{|f(z_0,t_0)|}^{|z_0|} \frac{4dx}{\sqrt{(1-x^2)}} \quad \left(|z_0| < \frac{1}{\sqrt{2}}\right),$$

$$\arg f'(z_0, t_0) = \int_{|f(z_0,t_0)|}^{\frac{1}{\sqrt{2}}} \frac{4dx}{\sqrt{(1-x^2)}} + \log \frac{|z_0|^2}{1-|z_0|^2} \quad \left(|z_0| > \frac{1}{\sqrt{2}}\right),$$

if $|f(z_0,t_0)| < \frac{1}{\sqrt{2}}$, and so for all large t_0. Thus the upper bounds of Theorem 6.6 may be approached as closely as we please and so are sharp.

6.13. Conclusion. The foregoing theorems represent some of the principal successes achieved by Löwner and his successors by means of Theorem 6.1. For many years this method was the only one capable of producing results of such depth. Recently two other methods have appeared which are capable of giving comparable results. These are Schiffer's variational method[†] and Jenkins's theory of modules, which was referred to at the end of Chapter 5. Both these methods can be used to prove $|a_3| \leqslant 3$.[‡] In addition both methods have led to a variety of results which go beyond this chapter. Thus by combining variational techniques with Theorem 6.1, Garabedian and Schiffer have been able to prove $|a_4| \leqslant 4$.[§] Unfortunately space has not permitted us to describe these difficult results in more detail and the reader must be referred to the original papers.

† Perhaps Schiffer[1] provides the best introduction.
‡ Jenkins[3], Garabedian and Schiffer[1].
§ Garabedian and Schiffer[2].

BIBLIOGRAPHY

(The following list consists only of those works which are quoted in the text.)

AHLFORS, L. [1] Untersuchungen zur Theorie der konformen Abbildung und der ganzen Funktionen. *Acta Soc. Sci. fenn.* series A, **1** (1930), no. 9.

[2] *Complex Analysis* (New York, 1953).

BAZILEVIČ, I. E. [1] On distortion theorems and coefficients of univalent functions. (Russian), *Mat. Sborn., N.S.* **28** (70), (1951), 147–64.

BIEBERBACH, L. [1] Über die Koeffizienten derjenigen Potenzreihen, welche eine schlichte Abbildung des Einheitskreises vermitteln. *S.B. preuss. Akad. Wiss.* **138** (1916), 940–55.

BIERNACKI, M. [1] Sur les fonctions multivalentes d'ordre p. *C.R. Acad. Sci., Paris*, **203** (1936), 449–51.

[2] Sur les fonctions en moyenne multivalentes. *Bull. Sci. Math.* (2), **70** (1946), 51–76.

BOHR, H. [1] Über einen Satz von Edmund Landau. *Scr. Bibl. Univ. Hierosolym*, **1**, no. 2 (1923).

BURKILL, J. C. [1] *The Lebesgue Integral* (Cambridge, 1951).

CARTWRIGHT, M. L. [1] Some inequalities in the theory of functions. *Math. Ann.* **111** (1935), 98–118.

DIEUDONNÉ, J. [1] Sur les fonctions univalentes. *C.R. Acad. Sci., Paris*, **192** (1931), 1148–50.

DVORETZKY, A. [1] Bounds for the coefficients of univalent functions. *Proc. Amer. Math. Soc.* **1** (1950), 629–35.

FEKETE, M. and SZEGÖ, G. [1] Eine Bemerkung über ungerade schlichte Funktionen. *J. Lond. Math. Soc.* **8** (1933), 85–9.

GARABEDIAN, P. R. and ROYDEN, H. A. L. [1] The one-quarter theorem for mean univalent functions. *Ann. Math.* (2), **59** (1954), 316–24.

GARABEDIAN, P. R. and SCHIFFER, M. [1] A coefficient inequality for schlicht functions. *Ann. Math.* (2), **61** (1955), 116–36.

[2] A proof of the Bieberbach conjecture for the fourth coefficient. *J. Rat. Mech. Anal.* **4** (1955), 427–65.

GOLUSIN, G. M. [1] On distortion theorems of schlicht conformal mappings. (Russian, German summary.) *Rec. Math., Moscou* (2), **1** (1936), 127–35.

[2] *Interior Problems of the Theory of Schlicht Functions*, translated by T. C. Doyle, A. C. Schaeffer, and D. C. Spencer (Office of Naval Research, Washington, 1947).

GRONWALL, T. H. [1] Sur la déformation dans la représentation conforme. *C.R. Acad. Sci., Paris*, **162** (1916), 249–52.

GRUNSKY, H. [1] Neue Abschätzungen zur konformen Abbildung ein- und mehrfach zusammenhängender Bereiche. *Schr. Inst. angew. Math. Univ. Berl.* **1** (1932), 95–140.

[2] Zwei Bemerkungen zur konformen Abbildung. *Jber. dtsch. Math. Ver.* **43** (1933), 1. Abt. pp. 140–3.

HARDY, G. H. [1] The mean value of the modulus of an analytic function. *Proc. Lond. Math. Soc.* (2), **14** (1915), 269–77.

HAYMAN, W. K. [1] *Symmetrization in the Theory of Functions.* Tech. Rep. no. 11, Navy Contract N 6-ori-106 Task Order 5 (Stanford University, 1950).

[2] Some applications of the transfinite diameter to the theory of functions. *J. Anal. Math.* **1** (1951), 155–79.

[3] Functions with values in a given domain. *Proc. Amer. Math. Soc.* **3** (1952), 428–32.

[4] The asymptotic behaviour of p-valent functions. *Proc. Lond. Math. Soc.* (3), **5** (1955), 257–84.

JENKINS, J. A. [1] On a problem of Gronwall. *Ann. Math.* (2), **59** (1954), 490–504.

[2] Some uniqueness results in the theory of symmetrization. *Ann. Math.* (2), **61** (1955), 106–15.

[3] On circumferentially mean p-valent functions. *Trans. Amer. Math. Soc.* **79** (1955), 423–8.

KAPLAN, W. [1] Close-to-convex schlicht functions. *Michigan Math. J.* **1** (1953), 169–85.

KOEBE, P. [1] Über die Uniformisierung der algebraischen Kurven. II. *Math. Ann.* **69** (1910), 1–81.

LANDAU, E. [1] Zum Koebeschen Verzerrungssatz. *R.C. Circ. mat. Palermo*, **46** (1922), 347–8.

LELONG-FERRAND, J. [1] *Représentation conforme et transformations à intégrale de Dirichlet bornée* (Paris, 1955).

LITTLEWOOD, J. E. [1] On inequalities in the theory of functions. *Proc. Lond. Math. Soc.* (2), **23** (1925), 481–519.

[2] On the coefficients of schlicht functions. *Quart. J. Math.* **9** (1938), 14–20.

[3] *Lectures on the Theory of Functions* (Oxford, 1944).

LITTLEWOOD, J. E. and PALEY, R. E. A. C. [1] A proof that an odd schlicht function has bounded coefficients. *J. Lond. Math. Soc.* **7** (1932), 167–9.

LÖWNER, K. [1] Untersuchungen über die Verzerrung bei konformen Abbildungen des Einheitskreises $|z| < 1$, die durch Funktionen mit nicht verschwindender Ableitung geliefert werden. *Leipzig Ber.* **69** (1917), 89–106.

[2] Untersuchungen über schlichte konforme Abbildungen des Einheitskreises. I. *Math. Ann.* **89** (1923), 103–21.

MONTEL, P. [1] *Leçons sur les fonctions univalentes ou multivalentes* (Paris, 1933).

NEVANLINNA, R. [1] Über die konforme Abbildung von Sterngebieten. *Öfvers. Finska Vet. Soc. Förh.* **53** (A) (1921), Nr. 6.

PÓLYA, G. [1] Sur la symétrisation circulaire. *C.R. Acad. Sci., Paris,* **230** (1950), 25–7.

PÓLYA, G. and SZEGÖ, G. [1] *Isoperimetric Inequalities in Mathematical Physics* (Princeton, 1951).

READE, M. O. [1] On close-to-convex univalent functions. *Michigan Math. J.* **3** (1955), 59–62.

ROBERTSON, M. S. [1] Multivalent functions of order *p*. *Bull. Amer. Math. Soc.* **44** (1938), 282–5.

ROGOSINSKI, W. W. [1] Über positive harmonische Sinusentwicklungen. *Jber. dtsch. Math. Ver.* **40** (1931), 2. Abt. pp. 33–5.

SCHAEFFER, A. C. and SPENCER, D. C. [1] The coefficients of schlicht functions. *Duke Math. J.* **10** (1943), 611–35.

[2] *Coefficient Regions for Schlicht Functions* (New York, 1950).

SCHIFFER, M. [1] Variation of the Green function and theory of the *p*-valued functions. *Amer. J. Math.* **65** (1943), 341–360.

SPENCER, D. C. [1] Note on some function-theoretic identities. *J. Lond. Math. Soc.* **15** (1940), 84–6.

[2] On finitely mean valent functions. *Proc. Lond. Math. Soc.* (2), **47** (1941), 201–11.

[3] On finitely mean valent functions. II. *Trans. Amer. Math. Soc.* **48** (1940), 418–35.

[4] On mean one-valent functions. *Ann. Math.* (2), **42** (1941), 614–33.

STEIN, P. [1] On a theorem of M. Riesz. *J. Lond. Math. Soc.* **8** (1933), 242–7.

SZÁSZ, O. [1] Über Funktionen die den Einheitskreis schlicht abbilden. *Jber. dtsch. Math. Ver.* **42** (1932), 1. Abt. pp. 73–5.

SZEGÖ, G. [1] Zur Theorie der schlichten Abbildungen. *Math. Ann.* **100** (1928), 188–211.

TITCHMARSH, E. C. [1] *The Theory of Functions,* 2nd ed. (Oxford, 1939).

INDEX

This index includes references to proper names and to topics of major importance in the theory developed here, but it is by no means exhaustive, and in the case of symbols and terminology usually the only reference given is to the pages containing important definitions.

$A(p)$, $A(p,q)$, 25
a_2, 1–4, 7–8, 115–16, 136; a_3, 4, 115–16, 118, 134–6, 144; a_4, 4, 144; a_5, 115; a_{nk+1}, 52–3; $a_{nk+\nu}$, 114; a_{p+1}, 99; a_{p+2}, 116; a_{p+k}, 116; a_{p+nk}, 114; arg a_n, 116
α, 9–11, 37, 94, 99–116
α_n, 107
$\alpha(t)$, 126; $\alpha(\zeta)$, 34
admissible domain, 64
Ahlfors, 8, 12, 17, 18, 36, 58, 124
analytic domain, 59
areally mean p-valent function, 94
argument of a function, 118, 138–43
asymptotic expansions, 105–12
averaging assumptions, 18, 22–4, 94

B_f, 120
$B_{t't''}$, 128
b_n, 104–5, 109–12; b_m, 136–8
β, 119, 129
$\beta(t)$, 127
Bazilevič, 11
Bieberbach, 1, 4, 10, 88, 115
Biernacki, 42, 50, 94
Bohr, 92
boundary behaviour in conformal mapping, 118–19
bounded univalent function, 50
branch point, 65–6
Burkill, 20, 21, 60

C.A., 8, 12, 58–9, 63, 65, 72, 85, 96
capacity, 58, 65–8, 75–6, 82–3
Cartwright, 17, 18, 31
circumferentially mean p-valent function, 94
closure, 59
coefficients, of univalent functions, 10; of mean p-valent functions, 50; of circumferentially mean p-valent functions, 104

condenser, 65–7, 75–6
connectivity, 59
continuity, modulus of, 72–4
convex domain, 12
convex function, 45
convex univalent function, 12–13
correspondence of boundary points, 120

$\bar{D}$, 59; D^*, 70; D_f, 84
d, $d(f)$, 119
Δ_n, $\Delta_n(\epsilon)$, 107
δ, $\delta(f)$, 120
denseness of a subclass, 117
diameter of a set, 119
Dieudonné, 13
Dirichlet, integral, 64, 76; problem of, 62–7
distortion theorems, 4, 17
domain, 1
doubly connected domain, 59
Dvoretzky, 51

$f(z,t)$, 117, 130
$f_k(z)$, 52–3, 113–14
$f_\theta(z)$, 2, 4–6, 9
Fekete and Szegö, 115
functions, of maximal growth, 37, 100; with k-fold symmetry, 52, 113; without zeros, 21, 95; with zeros, 25, 98

$G(t)$, 126
$g(z, z_0, D)$, 78
$g_t(z)$, 127
$\Gamma(x)$, 104
γ_t, $\gamma_{t't''}$, 126
Garabedian and Royden, 99
Garabedian and Schiffer, 4, 144
Gauss's formulae, 60
Golusin, 142
Green's, formula, 62; function, 78

Gronwall, 4, 141
Grunsky, 139, 142

$H(R)$, 22
$h(R)$, 22
$h(z, t', t'')$, 127
Hardy, 42
harmonic function, 62–5, 78–81
Hayman, 51, 58, 80, 84, 85, 95, 99, 116

$I_1(r, f)$, 10–11; $I_\lambda(r, f)$, 42–8, 116
$I_D(u)$, 64–8, 75–8
inner radius, 58, 79–84
inverse function, 136

Jenkins, 82, 84, 115, 116, 144

$\kappa(t)$, 117, 130–40
Kaplan, 14
Koebe, 1, 88

$\lambda(t)$, 127–31
Landau, 88
Lebesgue integral, 18, 20
Lelong-Ferrand, 18
length-area principle, 18
Lip or Lipschitzian function, 59–62, 65–8, 72–3, 75, 132
Littlewood, 10, 41, 50
Littlewood and Paley, 41, 52
Löwner, 4, 12, 116, 117–44

$M(r, f)$, 9
μ_q, 17, 25, 31–2
maximum modulus, 9
mean p-valent function, 23–57, 94–116
means, 10, 42
minimum modulus, 26
modules, theory of, 144
modulus, of a doubly connected domain, 65; of continuity, 72–4

$n(w)$, $n(w, \Delta, f)$, 18, 94
$n(r, w)$, 42
Nevanlinna, 14

O, O^*, 68
$\Omega(\delta)$, 72–4
$\omega(z)$, 65–7, 75; $\omega^*(z)$, 75
odd univalent function, 115

omitted values, 51
order of a function at a boundary point, 34

$p(R)$, $p(r, R)$, $p(r, \Delta, f)$, 18, 42, 94
Pólya, 69
Pólya and Szegö, 58, 68, 75, 76, 81, 82
potential function, 65–7, 75
power series with gaps, 55
principal frequency, 58
Principle of Symmetrization, 84
Problem of Dirichlet, 62–7
p-valent function, 1, 17–8, 42, 50, 53, 58

R, R_f, 87, 90
r_0, 79; r_0^*, 82; r_c, 141; r_s, 142
radius, inner, 58, 79; of convexity, 118, 141; of greatest growth, 101; of starlikeness or starshapedness, 118, 142
Reade, 14
real coefficients, 14, 115
reflexion principle, 36
region, 59
regularity theorem, 99, 116
Riemann mapping theorem, 36, 85, 124
Robertson, 53
Rogosinski, 13, 85
Rouché's theorem, 8, 96

S_f, 119
$S_\lambda(r)$, 44
$S_{t't''}$, 128
$\mathfrak{S}$, 1; $\mathfrak{S}_0$, 6; $\mathfrak{S}''$, 124
s.a. cut, 124
Schaeffer and Spencer, 115
Schiffer, 144
Schwarz's inequality, 19; lemma, 12; reflexion principle, 36
sectionally analytic cut or slit, 117, 124
simply connected domain, 59
slit mappings, 124
Spencer, 17, 18, 22, 31, 34, 42, 44, 50, 53, 99, 115
starlike or starshaped, domain, 14, 141; univalent function, 14–16; see also radius
Stein, 42
Steiner, 68

symmetrization, 68–93; Pólya, 69; Steiner, 68; of condensers, 75; of functions, 72
Szász, 13
Szegö, 4

$T(r)$, 90–1
τ, 129
Titchmarsh, 19, 104, 109
torsional rigidity, 58
transfinite diameter, 58

transformation, 119–29
typically real function, 13–6

$u^*(x)$, 72
univalent function, 1

variational method, 144
Vitali's convergence theorem, 109

$W(R)$, 22

PUBLISHED BY
THE SYNDICS OF THE CAMBRIDGE UNIVERSITY PRESS
Bentley House, 200 Euston Road, London, N.W. 1
American Branch: 32 East 57th Street, New York 22, N.Y.

PRINTED IN GREAT BRITAIN